SYSTEMS FOR CYTOGENETIC ANALYSIS IN *VICIA FABA* L.

ADVANCES IN AGRICULTURAL BIOTECHNOLOGY

Akazawa T., et al., eds: The New Frontiers in Plant Biochemistry. 1983.
ISBN 90-247-2829-0

Gottschalk W. and Müller H.P., eds: Seed Proteins: Biochemistry, Genetics, Nutritive Value. 1983. ISBN 90-247-2789-8

Marcelle R., Clijsters H. and Van Poucke M., eds: Effects of Stress on Photosynthesis. 1983. ISBN 90-247-2799-5

Veeger C. and Newton W.E., eds: Advances in Nitrogen Fixation Research. 1984.
ISBN 90-247-2906-8

Chinoy N.J., ed: The Role of Ascorbic Acid in Growth, Differentiation and Metabolism of Plants. 1984. ISBN 90-247-2908-4

Witcombe J.R. and Erskine W., eds: Genetic Resources and Their Exploitation – Chickpeas, Faba beans and Lentils. 1984. ISBN 90-247-2939-4

Sybesma C., ed: Advances in Photosynthesis Research. Vols. I-IV. 1984.
ISBN 90-247-2946-7

Sironval C., and Brouers M., eds: Protochlorophyllide Reduction and Greening.
1984. ISBN 90-247-2954-8

Fuchs Y., and Chalutz E., eds: Ethylene: Biochemical, Physiological and Applied Aspects. 1984. ISBN 90-247-2984-X

Collins G.B., and Petolino J.G., eds: Applications of Genetic Engineering to Crop Improvement. 1984. ISBN 90-247-3084-8

Chapman G.P., and Tarawali S.A., eds: Systems for Cytogenetic Analysis in *Vicia Faba* L. 1984. ISBN 90-247-3089-9

Systems for Cytogenetic Analysis in *Vicia Faba* L.

Proceedings of a Seminar in the EEC Programme of Coordination of Research on Plant Productivity, held at Wye College, 9–13 April 1984

Sponsored by the Commission of the European Communities, Directorate-General for Agriculture, Coordination of Agricultural Research

edited by

G.P. CHAPMAN
S.A. TARAWALI

Wye University College
University of London
United Kingdom

1984 **MARTINUS NIJHOFF/DR W. JUNK PUBLISHERS**
a member of the KLUWER ACADEMIC PUBLISHERS GROUP
DORDRECHT / BOSTON / LANCASTER
for
THE COMMISSION OF THE EUROPEAN COMMUNITIES

Distributors

for the United States and Canada: Kluwer Academic Publishers, 190 Old Derby Street, Hingham, MA 02043, USA

for the UK and Ireland: Kluwer Academic Publishers, MTP Press Limited, Falcon House, Queen Square, Lancaster LA1 1RN, England

for all other countries: Kluwer Academic Publishers Group, Distribution Center, P.O. Box 322, 3300 AH Dordrecht, The Netherlands

Library of Congress Cataloging in Publication Data

```
Main entry under title:

Systems for cytogenetic analysis in Vicia faba L.

   (Advances in agricultural biotechnology ; 11)
   "Sponsored by the Commission of the European
Communities, Directorate-General for Agriculture,
Coordination of Agricultural Research."
   Includes bibliographical references and index.
   1. Faba bean--Genetics--Congresses.  2. Faba bean--
Breeding--Congresses.  I. Chapman, G. P.  II. Tarawali,
S. A.  III. Commission of the European Communities.
Coordination of Agricultural Research.  IV. EEC
Programme of Coordination of Research on the Improve-

ment of the Production of Plant Proteins.  V. Series.
SB351.F3S97  1984      583'.322        84-20724
```

ISBN-13: 978-94-009-6212-5   e-ISBN-13: 978-94-009-6210-1
DOI: 10.1007/978-94-009-6210-1

Book information

Publication arranged by: Commission of the European Communities, Directorate-General Information Market and Innovation, Luxembourg

Copyright/legal notice

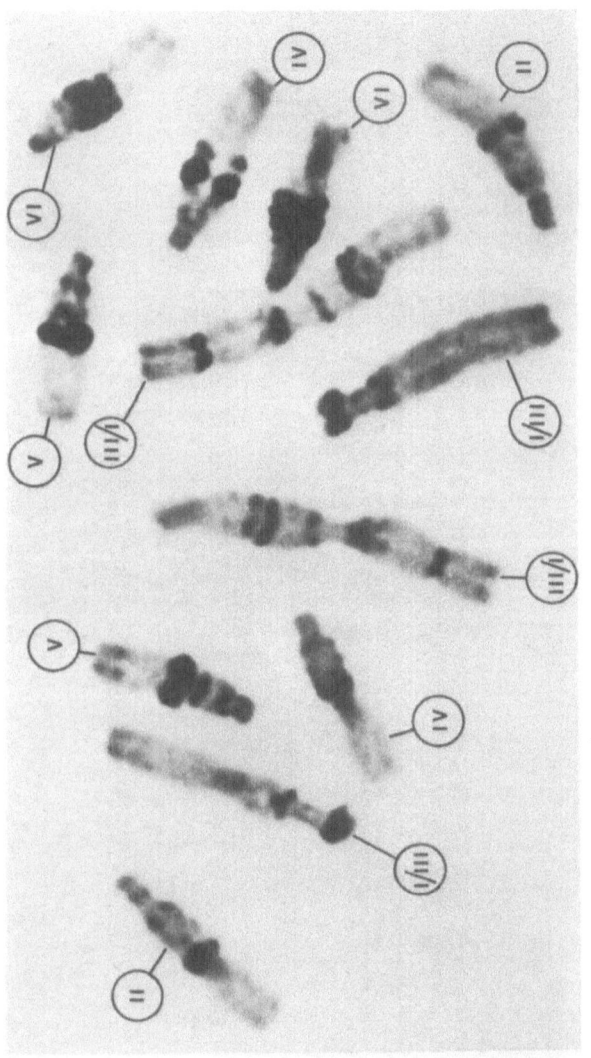

A Giemsa banded representation of the karyotype 'D' produced from an interchange between chromosomes I and III of the normal karyotype.

Photograph supplied by courtesy Professor R. Rieger

PREFACE

Unlike the situation in the major cereals, the yields of _Vicia faba_ have not been markedly increased during the last half century. There is no single cause for this but among those that have been important is the lack of cytogenetic understanding in relation to breeding performance. Since as a consequence, little genetic variation has been available to agronomists conclusions, probably unwarranted, have been drawn about the limited prospects for the faba bean. Against such a background it has been difficult to justify the investment of research resources in the crop.

The central theme of this book is that with the establishment of cytogenetic studies in _Vicia faba_ understanding of its genetic system will develop in relation to breeding improvement and thereafter some at least, of the impediments to yield increase can steadily though not dramatically, be removed.

We have distinguished between longer and shorter papers and only the former include Abstracts. The latter amplify themes in the longer papers or were written to develop particular topics at the request of the editors.

G.P. Chapman

S.A. Tarawali

Wye College, April, 1984

ACKNOWLEDGEMENTS

We would like to thank the various contributors to this publication for the readiness with which they have met our various requests.

Our thanks are due to the staff of the Centre for European Agricultural Studies for facilitating arrangements for the Seminar and to Mr. Peter Abott and Carl Zeiss (Oberkochen) Ltd. for the display of microscopes.

It is a pleasure too, to acknowledge the help of colleagues in the Department of Biological Sciences at Wye, especially Dr. J.W. Mansfield, Dr. W.E. Peat and Mrs. S. Elliss. We thank Dr. C.J. Hodgson for the drawing of Bombus hortorum on page 107.

Lastly, the editors would like to thank Mrs. S. Briant for her typing and her patience.

WELCOME TO DELEGATES BY THE PRINCIPAL OF WYE COLLEGE

I.A.M. Lucas C.B.E.

As an animal nutritionist my interest in Vicia faba is that it has more protein and more lysine in the protein than have the cereal grains. On the other hand, it needs to be combined with other feeds richer in methionine, it may have too small a proportion of threonine for some purposes and if given as over about 15% of the diet does not achieve its nutritional potential apparently because of the presence of tannin and possibly trypsin inhibitor and other substances. But there are varietal differences in chemical composition and this is where my interest reaches out towards your own.

Other members of Wye College look at Vicia faba from their own different points of view. Those concerned with third world development know of its role in human nutrition; plant pathologists are aware of how difficult breeding for resistance to diseases such as chocolate spot has proved to be. Agriculturalists recognize its value as a break crop in cereal rotations but economists know that it is not an especially favoured crop in EEC countries at present because of price structures. Librarians take into their collections the excellent reviews of Vicia faba which have been published in recent years. The crop is important and growing in importance world wide and it is evident that improvement in yield, nutritional value and reliability depend on the accumulation and application of fundamental knowledge of its biology.

Interest in chromosome structure, aberrations and manipulation and in gene linkage patterns has escalated and Wye College took the initiative in 1983 to organize the First International Vicia faba Cytogenetics Review Meeting. Now we have the Second Vicia faba Cytogenetics Review Meeting, with Dr. Chapman as the local organizer and I know that he hopes it will result in a book which will be not only a record of the discussions but also a practical manual that should in a very direct way advance the subject. This will be published under the auspices of the EEC. Priority areas for future research should become evident and I know that one of Dr. Chapman's enthusiasms is that in furthering these researches there should be a movement of young workers internationally between laboratories such as that already experienced between Wye and Naples. International co-operation and exchange would certainly be in accord with both EEC and world-wide philosophies.

In welcoming the Seminar on behalf of Wye College, I should give you a briefing on the College. It is in effect the 'Agricultural Faculty' of the Federal University of London, although strictly it is a part of the Faculty of Science. There are 560 students, about 30% of whom are postgraduates. Of the latter, about one half are from forty-one overseas countries. The subjects taught and researched range from crop and animal production, landscape design and environmental studies to economics and the applied natural sciences. Two departments are concerned only with research; the Hop Research Department and the Centre for European Agricultural Studies. There are 80 members of academic staff including 15 engaged wholly in research. All the others combine research with teaching.

The College has a long history, having been founded in 1447 as a secular establishment for priests but converted to a private residence 100 years later at the Reformation. It was not until 1894 that the counties of Kent and Surrey bought the original buildings to start agricultural teaching at the diploma level. By 1900 a link had been established with London University and in 1947, after much expansion, the College was granted a Royal Charter and became a School of the University; thereafter teaching only at degree level. There are some fine old buildings and gardens and I hope that members of the Seminar may find sufficient spare time in which to visit them. I wish you a successful meeting and a happy visit to Wye.

LIST OF PARTICIPANTS
(alphabetical)

Mr. P. Abbott	Carl Zeiss (Oberkochen) Ltd., P.O. Box 78, Woodfield Road, Welwyn Garden City, Hertfordshire AL7 1LU
*Prof. Dr. A.M.T. Abo-Hegazi	Radiobiology Dept. Atomic Energy Establishment, P.O. Box, Cairo, Egypt
Miss P.M. Allington	Biological Sciences Dept., Wye College Ashford, Kent. TN25 5AH, U.K.
Mr. M.P. Bailey	Biological Sciences Dept., Wye College Ashford, Kent. TN25 5AH, U.K.
Dr. D.A. Bond	Plant Breeding Institute, Maris Lane, Trumpington, Cambridge, U.K.
Miss L. Cardenas	Institute of Biological Sciences, University of Philippines, Los Banos College, Laguna, Philippines.
Dr. G.P. Chapman	Biological Sciences Dept., Wye College, Ashford, Kent. TN25 5AH, U.K.
Dr. E. Filippone	Centro di Studio per il Miglioramento Genetico Degli Ortaggi C.N.R., Via Universita, 133, Parco Gussone, Portici, Napoli, Italy
Mr. J. Higgins	National Institute of Agricultural Botany, Cambridge, CB3 0LE, U.K.
Dr. S.A. Khalil	Agricultural Research Center, Food Legume Research Section, Field Crop Institute, Giza, Egypt.
Dr. J.W. Mansfield	Biological Sciences Dept., Wye College, Ashford, Kent. TN25 5AH, U.K.
*Dr. A. Martin	Escuela Tecnica Superior de Ingenieros Agronomis, Departamento de Genetica, Apartado 3048, Cordoba, Spain.
Mr. M.R. Martin	Biological Sciences Dept., Wye College, Ashford, Kent. TN25 5AH, U.K.
Dr. R.A. Neve	Hop Research Department, Wye College, Ashford, Kent. TN25 5AH, U.K.
Dr. S.A. Omar	ICARDA, P.O. Box 4416, Cairo, Egypt.
*Dr. T. Paratasilpin	Plant Pathology Dept., Chiang Mai University, Chiang Mai, Thailand.

Dr. W.E. Peat

Biological Sciences Dept., Wye College,
Ashford, Kent. TN25 5AH, U.K.

Dr. B. Pickersgill

Dept. of Agricultural Botany, Plant
Sciences Laboratories, University of
Reading, Whiteknights, Reading, Berks.
RG6 2AS, U.K.

Dr. G. Ramsay

Dept. of Agricultural Botany, Plant
Sciences Laboratories, University of
Reading, Whiteknights, Reading, Berks.
RG6 2AS, U.K.

Prof. Dr. R. Rieger

Zentralinstitut fur Genetik und
Kulturpflanforschung, DDR 4325
Gatersleben, Correnstrasse 3

Dr. L.D. Robertson

ICARDA, P.O. Box 5466, Aleppo, Syria.

Dr. S.H. Salih

Hadeiba Research Station, Ed-Damer,
Sudan

Dr. S.A. Tarawali

Biological Sciences Dept., Wye College
Ashford, Kent. TN25 5AH. U.K.

Prof. K. Yamamoto

Faculty of Agriculture, Kagawa
University, Miko-Tyo Institution,
Kagawa-Ken, Japan.

*These papers were presented at the meeting in the absence of the
contributors

CONTENTS

XVI

VICIA FABA CYTOGENETICS AND BREEDING: AN INTRODUCTION

G.P. Chapman

Department of Biological Sciences,

Wye College, Ashford, Kent TN25 5AH, U.K.

In the major crop plants there is now emerging 'molecular cytogenetics'. This is not merely the association of a gene with a chromosome region but an ambitious and increasingly successful attempt to relate gene function to development. Where resources are more abundant, progress tends to be more rapid as in the cereals, for example.

BACKGROUND

The cytology of Vicia faba is of classical interest and has yielded such fundamental discoveries as that by Thoday and Read in 1947 of the 'oxygen effect' whereby radiation damage is intensified in the presence of that element. In 1957 Taylor and his associates used Vicia faba to demonstrate the first instance of semi-conservative replication in a eukaryote chromosome. Caspersson with his co-workers demonstrated in 1969 that quinacrine mustard compounds could induce fluorescent banding in the chromosomes of this species, a discovery rapidly applied subsequently, to chromosomes of human and other species. A relatively recent review of medically related studies is that due to Kihlman (1977) for this species.

More recently the 'ready made' cytology of Vicia faba has developed in new directions. Two in particular can be mentioned. The first has been to integrate it with the accumulating data from bean breeding to inaugurate the cytogenetic study of this species. The second has been the extension to plant chromosomes of ultrastructural studies pioneered with animal chromosomes. Those aspects increasingly interact and contribute toward an understanding of the genome with which the breeder works. For a detailed review of Vicia faba cytogenetics the reader is referred to Chapman (1983b).

BREEDING

During the last two decades breeding of Vicia faba has expanded substantially and with it alternative views of what is the most rapid means of genetic improvement. A comprehensive review is that of Lawes,

Bond and Poulsen (1983). Normally the plant is open pollinated and indeterminate and attempts have been made to convert it to an F_1 hybrid system or more radically to convert it to a 'cereal mimic' having determinate growth with a terminal inflorescence. Conversion too, to a closed flower that automatically self pollinates is possible and might be advantageous if it could be separated from the seemingly inevitable inbreeding depression,and acquire too, adequate autofertility. Among breeders opinions also differ about the relative merits of winter and spring beans and for an area the size of Europe, what would be the optimum number of regional varieties required. Regional testing has shown differences in yield stability for a range of environments (Dantuma, von Kittlitz, Frauen and Bond, 1983). Experimentally, the best open pollinated determinate types now virtually match the best indeterminates. If this were sustained then a concentration of resources might well shift toward the more convenient determinate and semi-determinate types and the familiar indeterminate growth habit could possibly be superseded.

Recent developments in the search for disease resistance genes, and those that control for example, closed flower, determinate growth and others have revealed the need for cytogenetic analysis of the faba genome.

Various practical difficulties impede progress here. Among geneticists working on the crop the proportion concerned with producing new varieties is probably too large compared with those asking more fundamental questions. Only recently has the need for widespread systematic collection of new germplasm been met. Again, in Europe there is a group of varieties not greatly dissimilar in performance whose status should probably be reduced merely to that of 'populations' and from which might then be withdrawn some of the resources these involve for more profitable use in newer directions such as more extensive and rigorous testing of many fewer but recognisably 'elite' materials.

Lastly, some breeding objectives such as the search for 'high protein' now require critical re-appraisal. Fortunately, the situation has begun to change and through the encouragement of the EEC and ICARDA there has been a co-ordination of scientific effort applied to virtually every aspect of the crop.

In 1983, the First International <u>Vicia</u> <u>faba</u> Cytogenetics Review Meeting was held at Wye College which brought together representatives of classical cytology, plant breeding, ultrastructure and molecular biology. No individual papers were published but a series of recommendations to standardise procedures and facilitate exchange of information were formally adopted, Chapman (1983a). Important among these were the adoption of a common chromosome numbering system, that of Michaelis and Rieger (1959, 1968), the decisions to accumulate a complete set of trisomics as part of a co-ordinated study among several institutions of linkage and to expand the list of available genetic variation and make interesting mutant forms available to breeders, physiologists, agronomists and pathologists. Arrangements were in the hands of a small co-ordinating committee. The list of genetic variation (Chapman, 1982) originally published by the International Centre for Agricultural Research in Dry Areas (ICARDA) is reprinted here by kind permission of that organisation as Appendix I with minor alterations only.

Routinely, advances in technique, new mutants and gene placement on chromosomes for example are reported in 'FABIS' (Faba Bean Information Service) and additionally, Faba Bean Abstracts are published by the Commonwealth Agricultural Bureaux.

THE CYTOGENETIC CONTRIBUTION

Several features of the Vicia faba genetic system are unusual, for example.

1. The few and large chromosomes form a characteristic karyotype typical of almost any population examined and the amount of DNA per nucleus exceeds that for the majority of diploid plants.

2. Experimentally, the karyotype can be 'remodelled' utilising translocations and inversions to give at least in isolation, stable new karyotypes that can be hybridised to yield predictable chromosome associations at meiosis. If however, the normal karyotype occurs in its tetraploid form it is, apparently, totally isolated sexually from its normal diploid counterpart.

3. Where for any major character alternative alleles are available these are concentrated at relatively few loci. None the less, chiasma formation occurs seemingly at random and no chromosome regions are excluded from recombination.

4. Populations normally consist of individuals that were derived from selfs that are more likely (though not invariably) to cross and form crosses that are more likely to self.

5. The species will not cross to yield germinable hybrids with any other species nor is any putative 'wild ancestor' recognisable.

The species therefore can be characterised as having, outside of special collections, populations with a common karyotype, widely distributed. These populations are intrinsically heterozygous but remain strictly isolated genetically from all other species. (By contrast, _Pisum sativum_ is highly self-pollinating but will cross with other _Pisum_ species.)

The Second International _Vicia faba_ Cytogenetics Review Meeting took as its theme the search for systems to facilitate cytogenetic analysis. The theme was interpreted broadly and included both fundamental and applied aspects. This book will, it is hoped, advance the subject by both defining problems and indicating the means currently available to approach them whether at the cell level, that of the individual plant, the population or that of the wider relationships among the Vicieae.

INTACT AND BROKEN CHROMOSOMES

The focus of interest is the normal karyotype since it is the basis of existing varieties and chromosome breakage giving remodelled karyotypes is of interest here in contributing to an understanding of the normal karyotype. Using this approach, Rieger and his colleagues have made a series of elegant studies on mutation sensitivity. Essentially, movement of chromosome regions around the karyotype can affect their sensitivity to mutagens. Regions especially breakage-prone, the so-called 'hot spots' in one location can sometimes become quiescent if shifted elsewhere. In a further development reported here Rieger (p. 40) demonstrates the effect of differential mutagen dosage to explore the likelihood of inducible repair processes. These results using detached root tip meristems now await modified application to intact shoot meristems as a possible means of generating useful new variation.

Giemsa banding (and indeed most other banding techniques) has two aspects. The first is a largely pragmatic one wherebt one chromosome in a karyptype can be reliably distinguished from another. The extension of this technique to a series of related species broadens the scope of enquiry in that it can help identify chromosome alterations (inversions, translocations and so on) perhaps associated with speciation but it additionally emphasises questions about the nature of chromosome banding. The paper by Ramsay explores within the genus Vicia some aspects of variation in staining procedures.

An alternative approach to chromosome breakage in _Vicia faba_ is that of Saccardo and Filippone where the consequences of a dicentric situation are described here in some detail. An added interest of this contribution is the inclusion of microdensitometric data.

Emphasis at Wye is on chromosome ultrastructure and the extent to which techniques for chromosome manipulation developed originally for human chromosomes can be applied to those of plants.

LINKAGE AND TRISOMICS

The construction of a linkage map has important practical consequences for crop plant improvement. Systematically it

i)   promotes a search for new and possibly useful variation

ii)   contributes to an understanding of genome structure and function

iii) as it becomes more detailed, offers the breeder more expeditious routes to preferred recombinations.

The peculiarities of the Vicia faba genome indicated earlier emphasise the need for a systematic approach.

Pioneer studies of linkage were made by Sjodin (1971c) for this species and the present early state of linkage studies is indicated in Appendix I.

The lower the haploid number, the simpler, theoretically, it is to construct a linkage map. For Vicia faba where n = 6 two major technical difficulties are first, the relative similarity in appearance of the five acrocentrics and second, the absence of trisomic stocks. For the first of these problems Giemsa banding is, in the hands of experts, useful to identify each chromosome type but conditions have to be very carefully controlled and a more satisfactory alternative is required. The absence of trisomics is now, due to the painstaking work of Martin and his collaborators on the way to being solved and the latest advances are reported in this volume. Given a complete set of trisomics, if it becomes possible for any sharply defined mutant to associate its departure from normal Mendelian segregation with the presence of one of the six possible trisomics it can then provisionally, be attributed to the corresponding linkage group. At present it seems unlikely that the large satellite chromosome can be tolerated by the plant in a trisomic state, its effects being too disruptive.

(There is at this juncture, an intriguing problem. If a gene could not be associated via trisomic analysis with any acrocentric chromosome it could be inferred to be associated with the satellite chromosome - a

manifestly inconclusive and unsatisfactory procedure. What is clearly
required is in regard to this chromosome, an alternative procedure giving
a more positive conclusion about these genes genuinely sited on it.)

In theory, three methods of generating trisomics have been considered
for faba beans. These are

i)   Crossing tetraploids with diploids and backcrossing the triploid to
the diploid. Despite repeated attempts this seemingly obvious
approach has been entirely unsuccessful. Some tetraploids however
are irregular at meiosis and yield aneuploid progeny among which
Martin (1978) found four trisomics.

ii)   Using two reconstructed karyotypes Schubert, Rieger and Michaelis
(1983) showed that meiosis in the hybrid involved a hexavalent
formation predictably. Among the possible gametes formed and that
survived into the next generation were those that by chance included
an extra chromosome due to the peculiarities of hexavalent
separation.

iii) By utilising an asynaptic mutant crossed with diploid plants
Gonzalez-Garcia and Martin (1983) obtained nine trisomics including
of course some duplicates. This worker considers that these together
with two persisting from i) above probably comprise the complete
acrocentric set, a result to be confirmed by C-banding.

Trisomic stocks are delicate and need great care in maintenance and
handling. It is not established but is likely, that the extra chromosomes
are more easily transmitted through the embryo sac though even here
transmissibility could be insufficient or differ among the various
trisomics. If transmissibility were insufficient, the backcross to the
double recessive would depart insufficiently from a normal segregation and
mislead the investigator.

DISEASE RESISTANCE

Remarkably little reliable data is available for the inheritance of
resistance to Botrytis fabae probably the most generally serious faba bean
disease. Recently, however, there have been important developments.
Systematic search for Botrytis resistance has located ILB 438 and its
derivative ILB 938 described by Robertson (p. 79). Three features are
especially interesting. It comes from the New World where faba bean
cultivation was introduced only about 400 years ago and not directly from
the Old World original centre of domestication. Secondly, this material

resists to some extent quite dissimilar diseases such as Botrytis and Uromyces shown here by Khalil. Thirdly, the Botrytis resistance is effective in wide range of environments with implications for avoiding perhaps a too great dependance on one source of resistance only.

Mansfield presents in this volume a molecular approach to the interaction of B. fabae and V. faba and it is currently an unexplored question as to how the phytoalexin patterns alter in the new sources of resistance.

Of both theoretical and practical interest is the study by Omar of interactions within V. faba of virus and fungus namely, antagonistic for virus together with rust and synergistic for virus together with chocolate spot. Perhaps central is the recognition that Botrytis resistance should be tested in part, in hosts that are virus infected.

POPULATION

Any 'variety' of Vicia faba is established and maintained as a population in which self and cross pollination both regularly occur. Although the species has been adapted to F_1 hybrid production experimentally, complete with cytoplasmic male sterility, maintainer and restorer systems, on a field scale, the system broke down.

Breeding is, in faba beans more than most species, essentially population enrichment. Chance cross pollination by insects and rogueing of 'off types' compete in the maintenance of 'purity'. What alternatives are there to secure more easily managed population structures? Three can be defined. One is to select for closed flower mutants that automatically self pollinate using a genetic background resistant to inbreeding if this can be found. Associated with this is the desirability of a diligent search for any small regional populations that naturally have a high degree of selfing with minimal inbreeding depression. Thirdly, one must enquire whether more fundamental reshaping of the genome is possible. Although unlikely to yield competitively, the narrow genetic base and total isolation from diploids makes the population structure of tetraploids of some theoretical interest in this regard.

The contribution here from Higgins and Evans examines the practicalities that confront European bean breeders. That from Peat and Yawooz is an ingenious use of isoenzymology to monitor non-destructively gene transfers within various populations and there is now a need to use this or some other technique to locate different population structures including for reasons indicated earlier, those that habitually inbreed.

RELATIONSHIPS OF VICIA FABA TO VICIA AND THE VICIEAE

With the possible exception of Cicer, the genera in Vicieae (Cicer, Lathyrus Lens, Pisum and Vicia) form a coherent grouping although no intergeneric hybrids are known. Where sought for, interspecies hybrids can in some cases be obtained. They are however the exceptions. No hybrids between Vicia faba and any other species produce germinable embryos and the genetic isolation of this species is a source of perennial fascination as succeeding generations of scientists seek to cross it with Vicia narbonensis - not a convincing wild ancestor but the most obvious (though not very close) relative. Since morphology and crossability support contrary viewpoints taxonomic re-arrangements have been proposed and these were recently reviewed by Cubero (1984) who argued convincingly for the faba bean to be left within Vicia but in a separate subgenus (Faba).

For any group of plants where sexual hybrids cannot be formed, recourse to protoplast fusion tends to overlook the likely genomic incompatibility that could ensue. A perceptive contribution to the debate is the paper by Zenkteler and Melchers (1978) where for non-leguminous genera they demonstrated the unpromising developmental results of wide crosses whether achieved by somatic or sexual hybridisation and there is in any case an alternative view of the limits of crossability. In Pisum for example, hybrids can be produced between P. sativum and other species including P. fulvum and where the performance of the embryo is associated partly with the direction of cross and partly with the degree of chromosome alteration between the two species. The point at which the embryo falters differs in various cases and offers a route to the molecular biology both of species divergence and of embryo differentiation, an approach described by Tarawali in this volume. Such an approach might eventually be applicable to the hybrids among V. narbonensis briefly described here by Yamamoto and to the early embryonic stages that have resulted from the sustained efforts of Ramsay and Pickersgill illustrated here of the early embryonic stages of V. faba x johannis hybrids.

HAPLOIDY

In an outbreeding species the availability of total homozygotes such as those theoretically available from doubled haploids could be useful in genetic analysis. Culturing pollen to yield haploids has not so far been especially rewarding among legumes. In section 6 is a summary of a detailed study by Paratasilpin showing what is presently possible for Vicia faba.

Whether or not there are as yet unsuspected fundamental constraints to haploid embryo production from legume pollen and whether the alternative route of unfertilised ovule culture would be more appropriate are now questions in need of detailed examination.

MUTANT PHYSIOLOGY

A single recessive gene will convert an indeterminate vegetative inflorescence into a determinate reproductive one. The physiological consequences for the plant involve a redistribution of activity. Leaves persist rather than senesce and tiller production is stimulated. The choice of genetic background is important and can render the plants precocious or late flowering or intermediate between these extremes.

A problem for the breeder is to decide if, by choice of appropriate background a particular mutant's expression can be enhanced sufficiently for incorporation into a new variety or whether it is intrinsically less productive - a question currently under review especially for determinate habit and closed flower mutants. Would it be possible for example to raise productivity per plant sufficiently with determinate growth and at the same time use higher plant densities so as to permit higher yield per unit area and be economically worth while.?

The study by Bailey summarised in section 7 is an example of how mutant physiology can be assessed and it should be emphasised that relative performance of the two growth habits could be varied due to (say) change of genetic background or of environment. See also Baker, Chapman, Standish and Bailey (1984). Lastly, the paper by Abo-Hegazi extends mutant physiology into what is manifestly one involving polygenic inheritance.

CONCLUDING COMMENTS

The most prevalent impression one has, especially of the earlier literature of Vicia faba, is the fragmentary and unrelated nature of the work. A worker in one country for example presents an analysis of yield components almost as if he were the first to do so, seemingly unaware of the host of similar studies that preceded his. Similar examples for other sorts of investigation could be found.

There is now especially therefore, a need for careful and critical synthesis of what is already known so as to avoid wasteful duplication of effort and to focus more clearly on important areas of uncertainty.

This collection of papers was assembled to meet part of this need for synthesis and it is too, an expression of the international collaboration that is now developing.

REFERENCES

Baker, D.A., Chapman, G.P., Standish, M. and Bailey, M.P. 1983. Assimilate partitioning in a determinate variety of field bean. pp. 191-200 in : Temperate Legumes, physiology, genetics and nodulation. Eds. Jones, D.G and Davies, D.R. Pitman, pp. 442.

Casperson, I., Zech, L., Modest, E.J., Foley, G.E., Wagh, N. and Simonsson, E. 1969. Chemical differentiation with fluorescent alkylating agents in Vicia faba metaphase chromosomes. Exp. Cell Res. 58 128-140

Chapman, G.P. 1981. Genetic variation in Vicia faba FABIS 3 Suppl. p. 1-12.

Chapman, G.P. 1983a. Report : 1st International Vicia faba Cytogenetics Review Meeting. FABIS. 6, 19-20

Chapman, G.P. 1983b. 'Cytogenetics' pp. 197-216 in The Faba Bean, a basis for improvement, Ed. P.D. Hebblethwaite. Pub. Butterworths, Lond. 573

Cubero, J.I. 1984. Taxonomy, distribution and evolution of the faba bean and its wild relatives. pp. 131-144. In Genetic Resources and their Exploitation - Chick peas, Faba Beans and Lentils. Eds. J.R. Witcombe and W. Erskine. Pub. Martinus Nijhoff, The Hague. pp. 256.

Dantuma, G., Kittlitz, E. von, Frauen, M. and Bond, D.A. 1983. Yield, yield stability and measurements of morphological and phenological characters of faba bean (Vicia faba L.) varieties grown in a wide range of environments in Western Europe.Zeitschrift fur Pflanzenzuchter 90 85-105.

Gonzalez-Garcia, J.A. and Martin, A. 1983. Development, use and handling of trisomics in Vicia faba L. FABIS 6, 10-11

Kihlman, B.A. 1977. Caffeine and Chromosomes. North Holland Publishing Co. Amsterdam, pp. 504

Lawes, D.A., Bond, D.A. and Poulsen, M.H. 1983. 'Classification, Origin, Breeding Methods and Objectives' in The Faba Bean. : a basis for improvement. Ed. P.D. Hebblethwaite, pub. Butterworths Lond. pp. 573.

Martin, A. 1978. Aneuploidy in Vicia faba L. J. Hered. 69, 421-423

Michaelis, A. and Rieger, R. 1959 Structurheterozygotie bei Vicia faba Zuchter 29, 354-361.

Michaelis, A. and Rieger, R. 1968. On the distribution between chromosomes of chemically induced chromatid aberrations: studies with a new karyotype of Vicia faba. Mutation Res. 6 81-92

Schubert, I., Rieger, R. and Michaelis, A. 1983. A method for directed production of definite aneuploids in Vicia faba L. FABIS 7 13-18

Sjodin, J. 1971c. Induced translocations in Vicia faba L. Hereditas. 68 1-34.

Taylor, J.H. 1957. The time and mode of duplication of chromosomes. Amer. Nat. 91 209-221.

Thoday, J.M. and Read, T.M. 1947. Effect of oxygen on the frequency of chromosome aberrations produced by X-rays. Nature. Lond. 160. 608

Zenkteler, M. and Melchers, G. 1978. In vitro hybridisation by sexual methods and by fusion of somatic protoplasts. Theor. Appl. Genet. 52 81-90

1
Intact and
Broken Chromosomes

CHROMOSOME MANIPULATION

S.A. Tarawali*, M.R. Martin and P.M. Allington

*Formerly Miss S.A. Cooke

Department of Biological Sciences, Wye College, University of London,
Ashford, Kent TN25 5AH, U.K.

ABSTRACT

For about a decade until 1980 new developments in cytology were largely those involving animal species. Recently attempts have been made to extend this area of enquiry to plant species and to develop a series of appropriately modified techniques for handling plant chromosomes. In this laboratory the chromosomes of Vicia faba have provided the starting material, the resulting techniques then being extended to chromosomes of other species, notably in the genera Pisum, Linum and Triticosecale.

INTRODUCTION

The chromosome is to the cytologist an organelle and to the biochemist a macromolecule, their perceptions converging at the electron microscope. 'Molecular cytogenetics' integrates data from DNA sequences, nucleic acid-protein associations (the nucleosomes) and finally their higher order assembly into structures recognisable with the light microscope (chromosomes). At each level of magnitude it is increasingly possible to probe the chromosome structure and function with restriction enzymes or by in situ hybridisation or even to isolate it altogether apparently functioning normally, from the cell matrix. Undoubtedly, the pace of enquiry has largely been set by investigators working with mammalian and dipteran chromosomes and when one seeks to apply the results of animal cytology to plant chromosomes such as Vicia faba the large size of the genome becomes apparent when compared with that for example in Homo sapiens.

While in higher eukaryotes, mitosis and meiosis each show an underlying consistency, interphase is less well understood. Nagl (1982) has shown that chromatin condensation patterns in animals are 'tissue specific' and in plants 'species specific'. Furthermore, DNA sequence exchange between chromosomes and mitochondria appears to have occurred in animal cells and those of non-green saprophytes. Chromosomes of green plants co-exist additionally with chloroplasts and sequence exchange appears possible here too (Ellis, 1982, 1983). To integrate optical and molecular studies of chromosomes, this Department began a study of chromosome ultrastructure, part of which this paper describes.

MATERIALS AND METHODS

1. Fixed chromosomes

a) Vicia faba. Chromosomes, from the standard or reconstructed 'E.F.' (Michaelis & Rieger, 1971) Vicia faba karyotype were prepared for transmission (TEM) or scanning electron microscopy (SEM) as described by Chapman and Cooke (1983).

For salt treatment isolated chromosomes were maintained for 2 hours at 22°C in chromosome isolation buffer (Chapman & Cooke, 1983) modified to contain 2.0 M sodium chloride. Following this, an equal volume of chromosome isolation buffer containing 0.01% Triton X-100 (B.D.H. Chemicals Ltd.) was added and the suspension centrifuged at 3000 rpm for 15 minutes. The majority of the supernatant was removed to leave a concentrated suspension of salt treated nuclei and chromosomes. For optical microscopy a drop of this suspension was mixed with a drop of 0.05% aqueous Toluidine Blue (Hopkin & Williams C.I.52040).

b) Other species. Isolated chromosomes from the flax species Linum grandiflorum "Rubrum" were prepared as follows.

Flax seeds were germinated on tissue paper moistened with distilled water at 22°C. After three days aqueous 0.05% colchicine (Sigma) was added for 3 hours the root tips were then removed and transferred to 2% formalin for 40 minutes at 4°C. Chromosome isolation and TEM were as described by Chapman and Cooke (ibid).

For staining with the fluorochrome DAPI (Sigma), a drop of the suspension of nuclei and chromosomes was air dried onto a numbered copper E.M. grid coated with carbon Formvar. The grid was preincubated in phosphate buffer pH 8.0 for 2 minutes, then incubated with 15 μl of DAPI at 1 μg/ml in phosphate buffer pH 8.0 for 1 hour. Unbound DAPI was washed off with excess phosphate buffer and the grid was not allowed to dry out prior to mounting in 15 μl phosphate buffer pH 8.0, 1:9, glycerol with phenylenediamine (Sigma. Johnson and Nogueira Araujo, 1981) at a final concentration of 1 μg/ml. The grid was then viewed using a Zeiss photomicroscope I equipped with an EPI Fluorescence Illuminator and filter set number 2.

Preliminary investigations indicate that it is possible to view chromosomes using TEM after fluorescence microscopy. The mountant was washed off with excess phosphate buffer and the grid air dried. The unmounted grid can then be examined using fluorescence microscopy followed by TEM. The chromosomes of air dried material show a loss of resolution in comparison to critically point dried material and studies are currently in progress to combine fluorescence with critical point drying.

2. Unfixed chromosomes

a) Normal mitosis. Unfixed Vicia faba nuclei were isolated as follows:

Approximately 100 colchicine treated root tips (0.05% aqueous colchicine at 22°C for 3 hours) were placed in the inverted lid of a plastic petri dish and rinsed well with buffer (modified from Griesbach et al., 1982: 15 mM HEPES (N-2-hydroxyethylpiperazine-N'-2-ethanesulfonic acid, Sigma, pH 7.2; 1 mM EDTA (ethylenediamine-1-tetracetic acid, Sigma); 0.5 mM spermine (Sigma); 0.5 mM spermidine (Sigma); 80 mM potassium chloride (B.D.H. Chemicals Ltd.); 20 mM sodium chloride (B.D.H. Chemicals Ltd.); 1% hexylene glycol (2-methyl-2,4-pentandiol, Fluka AG)). The roots, in a few drops of buffer were coarsely chopped with a razor blade, covered with fine nylon gauze and squashed using the petri dish base. The resultant suspension was filtered with buffer through two layers of Miracloth (Calbiochem-Behring Corp.). The suspension was centrifuged at 800 rpm for 10 minutes, most of the supernatant was removed to leave a concentrated suspension of nuclei in buffer. The effects of sodium chloride on isolated, unfixed nuclei were studied using phase contrast microscopy. A drop of suspension was placed in buffer containing 0.25 M sodium chloride and added dropwise at one edge of the coverslip and a filter paper wick applied to the opposite edge to draw the solution through. The sodium chloride was removed by adding the original buffer (with no sodium chloride) in a similar manner.

b) Polyteny. Polyteny was induced in mature pea cotyledons by culture in a liquid medium containing 2,4-D as described by Davis & Cullis (1982). Details of the method used in this laboratory are described by Tarawali in this volume (p.132).

RESULTS

1. Fixed chromosomes

a) Vicia faba. Normal fixation using carnoy fluid or 3:1 absolute alcohol:acetic acid removes up to 80 per cent of chromosome proteins (Reteif & Ruchel, 1977) and retains the chromosomes within the cytoplasm surrounded by a cell wall. These latter components comprise debris that impedes resolution with the optical microscope. McLeish (1963) showed that a combination of brief formalin fixation and physical pressure could spring nuclei from within the cell. Chapman and Cooke (ibid) presented an adaptation of this to study isolated critical point dried chromosomes using the transmission electron microscope (Fig. 1). Isolated chromosomes on a coverslip can be gold coated and viewed using the scanning electron microscope (Fig. 2)

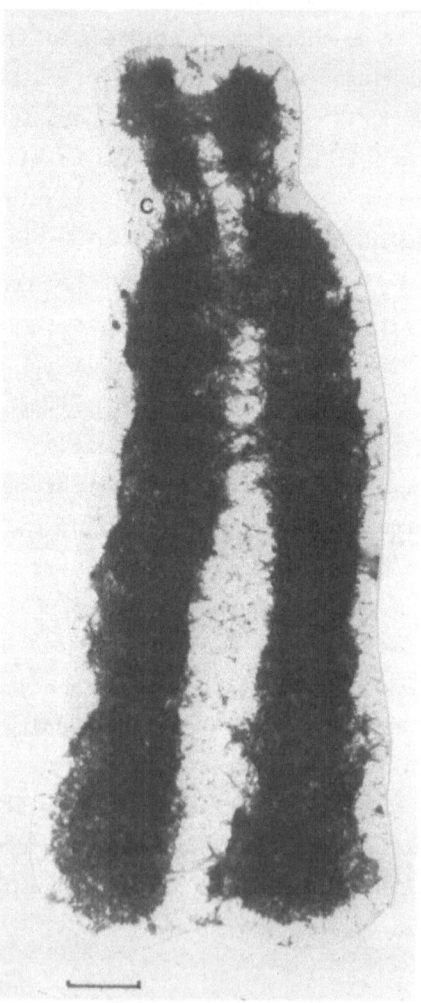

Figure 1.   Formalin fixed isolated, critical point dried V. faba
    chromosome number 2 (standard karyotype) using transmission
    electron microscopy. The fibrous nature of the chromosome is
    evident, although fibres are not packed at equal density along
    the chromatid length. C = centromere
    Bar = 1.0 μm

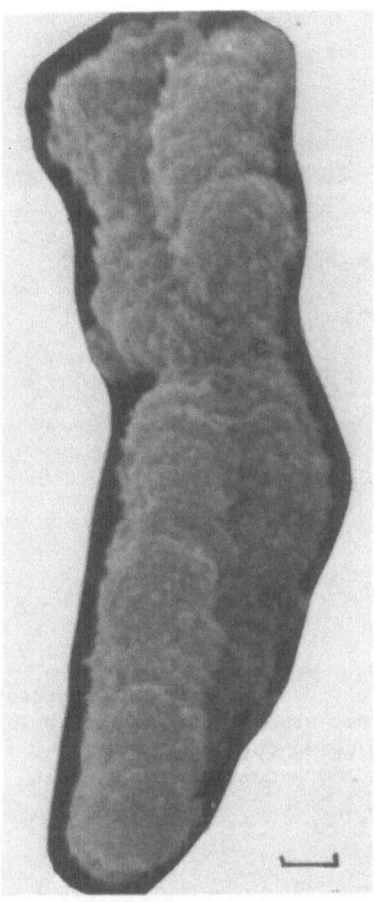

Figure 2. Formalin fixed, isolated, critical point dried
V. faba chromosome number 3 (EF karyotype), coated
with gold and viewed using scanning electron microscopy.
C = centromere.
Bar = 0.5 μm

A further modification is possible. For light microscopy, formalin
treated chromosomes retain a residual responsiveness to salt treatment
which induces a differential decontraction. The significance is not
understood but the patterns differ for each chromosome - presumably
reflecting underlying differences in chromatin fibre arrangement (Fig. 3)

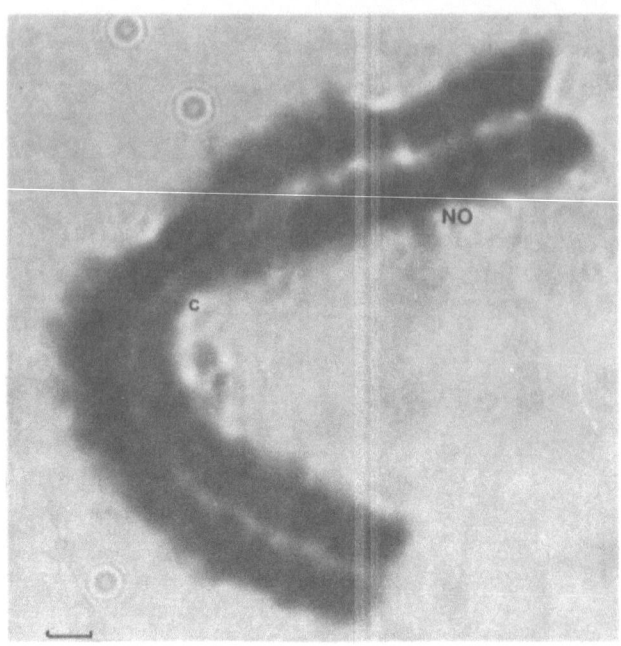

Figure 3. Formalin fixed, isolated, salt treated (see text)
V. faba chromosome number 1 stained with Toluidine
Blue and viewed using optical microscopy.
C = centromere; NO = nucleolus organiser
Bar = 2.0 μm

Fixed chromosomes are therefore relatively robust structures that can
be isolated, centrifuged, salt treated or critical point dried and gold
coated and retain their individuality throughout. What is unclear is the
degree of the structural deterioration imposed by formalin as a result of,
or in addition to, cross linking chromatin.

b) Adaptation to other species. Work in this laboratory has shown
that formalin fixed chromosomes of other Vicia species and those of
barley respond similarly to those of Vicia faba and a detailed study by
Martin (unpublished) working with the very small chromosomes of the flax
species Linum grandiflorum "rubrum" has shown even here that the
technique can be used to deposit and locate chromosomes on E.M. grids.
Since flax chromosomes are so small (seldom more than 2.0 μm in length
after their isolation) and thus difficult to separate from similar sized
cell debris, the search for chromosomes is extremely laborious and time-
consuming. Martin has shown that incorporation of the fluorochrome DAPI

into flax chromosomes on numbered EM grids enables their position in a particular square to be rapidly and precisely located with fluorescence microscopy. The prospects for subsequent search at the E.M. level appear greatly enhanced (see Figs. 4 and 5).

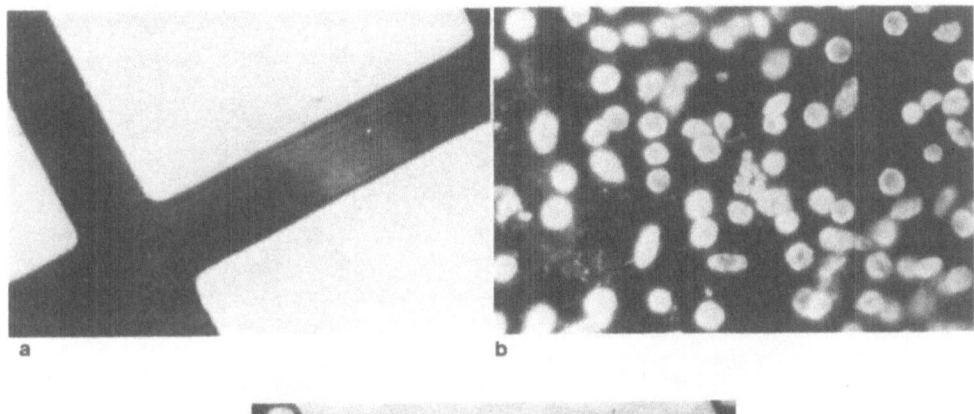

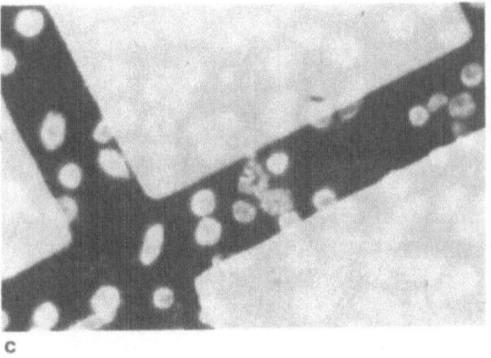

Figure 4. Deposited on an E.M. grid are DAPI stained L. grandiflorum nuclei and chromosomes. The same field of view is illustrated in all three cases.
a) Tungsten, bright field illumination only
b) Ultraviolet, fluorescent illumination only
c) Bright field and fluorescent illumination
Bar = 20 μm

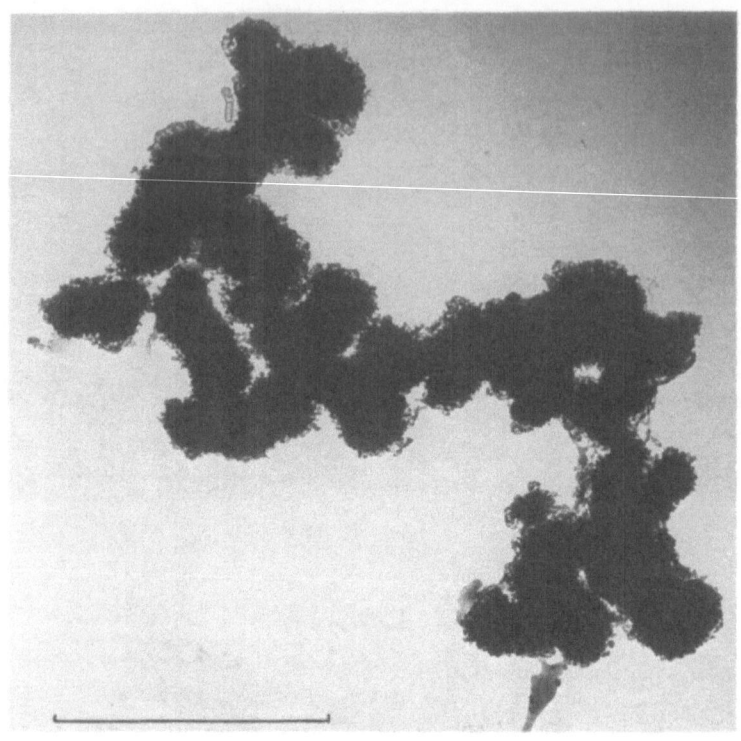

Figure 5. Formalin fixed, isolated, critical point dried
L. grandiflorum chromosome group viewed using
transmission electron microscopy.
Bar = 1.0 μm

c) Implications for Vicia faba. Hitherto the E.M. studies of Vicia
faba chromosomes have relied on the utilisation of the distinctive
remodelled karyotypes obtained by Michaelis & Rieger (1971) where the
relative size and form of each chromosome made it distinctive in the
absence of any staining procedure.

It has long been known since the work of Caspersson et al.(1969) that
each chromosome in the Vicia faba karyotype has a distinctive fluorescent
banding pattern. Were it to be possible (as seems likely) that critical
point dried chromosomes on E.M. grids could be induced to band-fluoresce
it would then be possible using a modification of Martin's procedure to
examine chromosomes interchangeably between light and transmission
microscopy.

There would be three significant gains.

1) It would be possible to utilise for E.M. studies the normal karyotype
   - the one of greatest interest to plant breeders.

2) It would be possible to obtain not representative chromosome types but
   individual chromosomes of a particular type by both optical and
   electron microscopy.

3) There is the intriguing prospect of being able to compare at the E.M.
   level, banded and unbanded chromosome regions.

In concluding this section it is worth stressing that where
chromosomes are large and dense compared with typical cell debris, careful
centrifuging renders them 'clean' and this combined with the absence of a
cell wall improves the clarity of even conventional light microscopy.

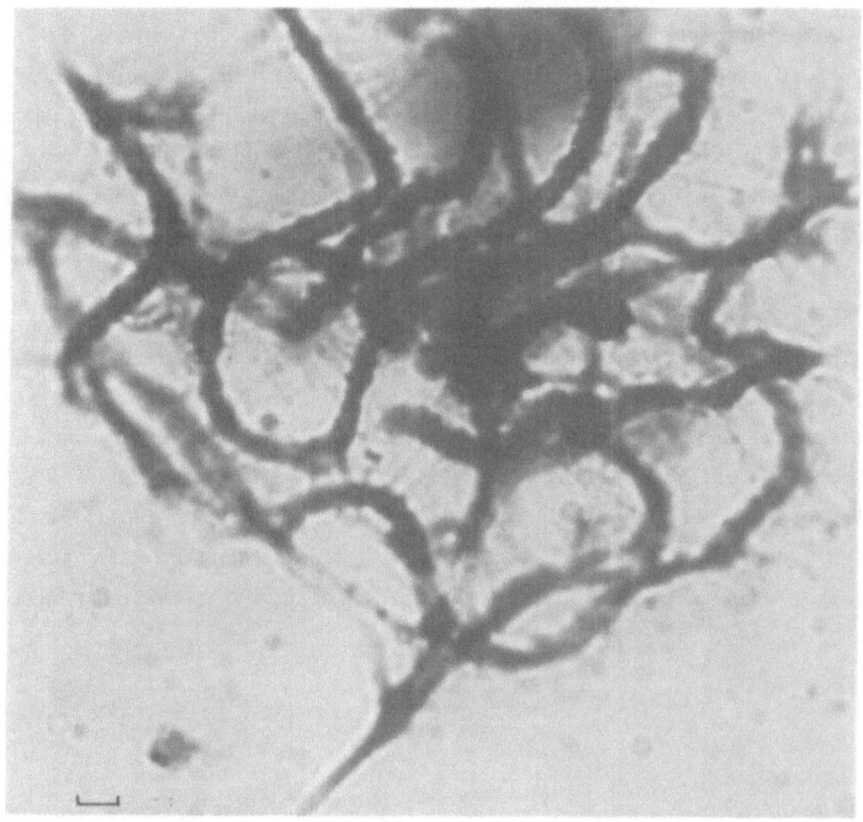

Figure 6.   Isolated, unfixed,V. faba prophase nucleus
    stained with Toluidine Blue and viewed using optical
    microscopy. Arrows indicate matching chromatid pairs.
    Bar = 2.0 μm

2. Unfixed chromosomes

    a)   Normal mitosis. Elsewhere other workers, notably Hadlaczky et al. (1983), Griesbach et al (1982), Malmberg et al. (1980) and Matthews (1983) have shown that the isolation of unfixed metaphase chromosomes of plants is possible. Some of the implications were examined recently by Chapman (in press).

In this laboratory unfixed nuclei at various stages from interphase to late prophase have been obtained (Fig 6) and it is possible to show the rapidly reversible dispersion and recompaction that can be induced by the addition and removal of sodium chloride from the buffer (demonstrated at the Wye Seminar using a video film prepared by Mrs. S.A. Tarawali).

It is noteworthy that chromosomes always re-compact to whatever stage of division they were at rather than to an earlier or later one.

b) Polyteny. In 1982 Davies and Cullis showed that polyteny could be readily induced in cultured pea cotyledons in culture, see Figs. 7 and 8. This was confirmed at Wye and one of us (P.M. Allington) found it possible to either (a) extract fixed cell-free polytene chromosomes or (b) extract unfixed polytene nuclei into buffer where they can be handled by a micromanipulator. (Work on pea cotyledons is considered in more detail on p. 134). Preliminary work with Vicia faba cotyledons suggests that polyteny is also inducible in that species. (For convenience, note that both fixed and unfixed aspects of polyteny are considered in this section.)

The presumed advantage of polytene chromosomes (hitherto not readily available in large quantities from plant tissues) lies in lateral replication thus multiplying in a conveniently orderly way the number of gene copies within one cell. Using this approach for salivary chromosomes in Drosophila, Wu and Davidson (1981) probed for single copy genes. So far, it does not appear that plant polytenes 'band' as do those of salivary glands and techniques for handling the plant equivalents have yet to be developed.

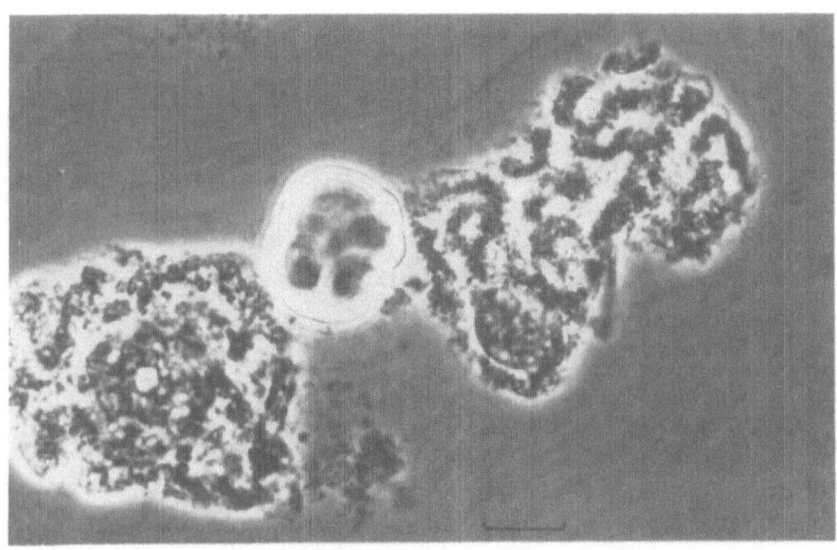

Figure 7. P. sativum polytene nuclei isolated from cultured
cotyledon tissue squashed in acetic acid and viewed using phase
contrast microscopy.
Bar = 10 µm.

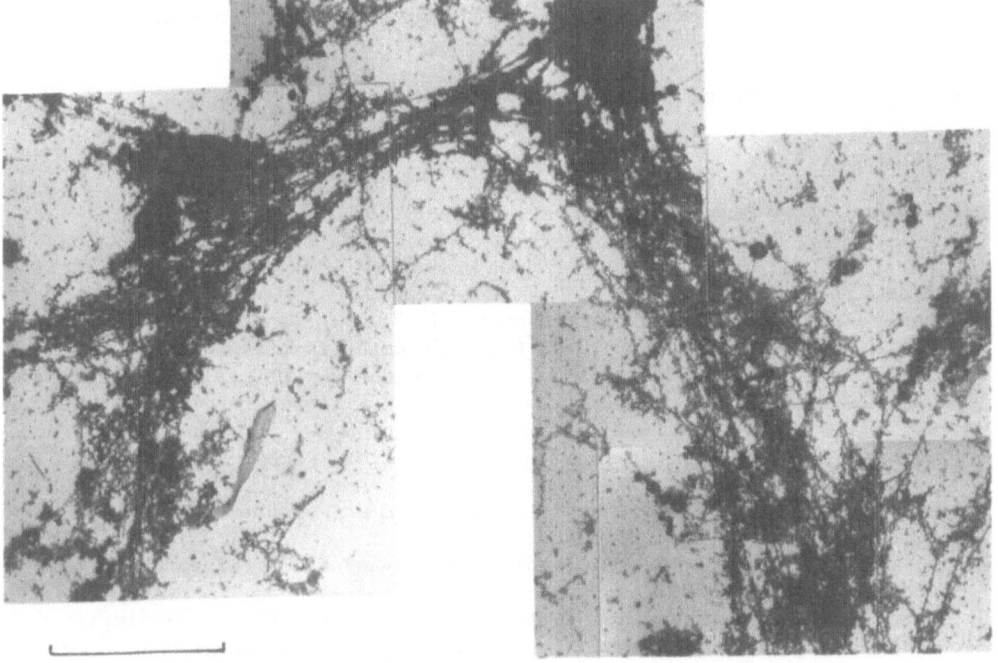

Figure 8. Unfixed polytene chromosome isolated from P. sativum
cotyledon culture. Viewed using the transmission electron
microscope after critical point drying. Bar = 5.0 µm

CONCLUSIONS

The few and large chromosomes of <u>Vicia faba</u> have made it a classic cytological subject. Emphasis is now changing from cytology to cytogenetics where various chromosome aberrations are not discarded but retained as informative parts of still-viable genomes. The existence of a 'ready-made' traditional cytology and the facility with which its chromosomes can now be manipulated both in the laboratory and as later authors will show, within living systems and the increasing choice of marker genes together comprise an promising area of enquiry. Additionally, the accumulation of 'gene libraries' such as those for pea (with obvious applicability to faba bean) the capacity to induce polyteny and the possibility of combining with probe nucleic acid both fluorescent and electron dense markers such as colloidal gold offers within about five years the prospect of biochemical linkage mapping for <u>Vicia faba</u> given the necessary resources.

One of the tasks of this meeting having identified this kind of objective, is to contrive the means to realise it. In doing so we involve two other groups of colleagues. These are firstly the breeders concerned with the practical improvement of the crop and secondly that much larger group for whom not only the cytology but also the cytogenetics of <u>Vicia faba</u> might now begin to constitute a 'model system'.

REFERENCES

Caspersson, T., Zech, L., Modest, E.J., Foley, G.E., Wagh, N. and Simonsson, E. 1969. Chemical differentiation with fluorescent alklylating agents in <u>Vicia faba</u> metaphase chromosomes. Exp. Cell Res. <u>58</u>, 128-140.

Chapman, G.P. The Evolved Chromosomes of Higher Plants. In : Plant Chromosome Ultrastructure, ed. G.P. Chapman, in press.

Chapman, G.P. and Cooke, S.A. 1983. A technique for optical and electron microscopy of isolated plant chromosomes. Protoplasma <u>116</u>, 198-200.

Davies, D.R. and Cullis, C.A. 1982. A simple plant polytene chromosome system and its use for <u>in situ</u> hybridisation. Plant Molec. Biol. <u>1</u>, 301-304.

Ellis, J. 1982. Promiscuous DNA-chloroplast genes inside plant mitochondria. Nature <u>299</u>, 678.

Ellis, J. 1983. Mobile genes of chloroplasts and the promiscuity of DNA. Nature, <u>304</u>, 308-309.

Griesbach, R.J., Malmberg, R.L. and Carlson, P.S. 1982. An improved technique for the isolation of higher plant chromosomes. Plant Sci. Lett. <u>24</u>, 55-60.

Hadlaczky, Gy, Bisztray, Gy, Praznovszky, T. and Dudits, T. 1983. Mass isolation of plant chromosomes and nuclei. Planta, <u>157</u>, 278-285.

Johnson, G.D. and Nogueira Araujo, G.M. de C. 1981. A simple method of reducing the fading of immunofluorescence during microscopy. J. Immunol. Methods 43, 349-350.

McLeish, J. 1963. Quantitative relationships between DNA and RNA in isolated plant nuclei. Proc. R. Soc. Ser. B. 158, 261-278.

Malmberg, R.L. and Griesbach, R.J. 1980. The isolation of mitotic and meiotic chromosomes from plant protoplasts Pl. Sci. Lett. 17, 141-147.

Matthews, B.F. 1983. Isolation of mitotic chromosomes from partially synchronised carrot, Daucus carota cell suspension cultures. Plant Sci. Lett. 31 165-172.

Michaelis, A. and Rieger, R. 1971. New karyotypes of Vicia faba. Chromosoma 35, 1-8.

Nagl, W. 1982. Condensed chromatin: species-specificity and cell cycle-specificity as monitored by scanning cytometry. In : Cell Growth, ed. C. Nicolini, Plenum Pub. Corp. pp. 171-218.

Reteif, A.E. and Ruchel, R. 1977. Histones removed by fixation. Exp. Cell Res. 106, 233-237.

Wu, M. and Davidson, N. 1981. Transmission E.M. method for gene mapping on polytene chromosomes by in situ hybridisation. Proc. Natl. Acad. Sci. 78, 7059-63.

C-BANDING IN VICIA SPECIES

Gavin Ramsay

Department of Agricultural Botany, Plant Science Laboratories
Whiteknights, University of Reading, Reading RG6 2AS, U.K.

ABSTRACT

Using Leishmans stain for C-banding, the authors indicate the use of this approach for identifying chromosome aberrations and clarifying taxonomic relationships and seek to apply them to resolve the origin and evolutionary history of V. faba. Chromosomes of six species of Vicia were examined including four accessions of V. faba. Details of banding are illustrated and two hypotheses considered to explain the evolution of the V. faba karyotype.

INTRODUCTION

C-banding techniques are used to reveal the patterns of constitutive heterochromatin along metaphase chromosomes. This constitutive heterochromatin represents genetically inert material containing repeated sequences of DNA which remains in a condensed form through interphase. In many organisms the C-bands are sufficiently distinctive in both number and position to allow the identification of individual chromosomes which may be indistinguishable by other means. This tool can be of use to those interested in crop plants in four main ways.

1. Genetic and chromosome studies. Chromosome variants, such as trisomics, other aneuploids, translocations, inversions, duplications and deletions can be identified using this technique (Dobel et al., 1973). If polymorphic C-bands occur, these can be used in gene mapping (e.g. barley, Linde-Laursen, 1979).

2. Intraspecific evolution. This may be investigated by comparing primitive and advanced types with each other and with related species, In Vicia faba L. for example, the small-seeded types, particularly var. paucijuga, are thought to be primitive so comparison with large-seeded types may indicate whether evolution under domestication has been accompanied by chromosomal structural rearrangement and/or gain or less of heterochromatin.

3. Cytotaxonomy. Where the relationships between species are disputed, for example in Vicia section Faba, C-banding may help clarify these

relationships and delimit taxonomic boundaries. Lavania and Sharma (1983) have reviewed the literature on the use of C-band patterns in interspecific comparisons in 17 higher plant genera.

4. Genome relationships. In addition to purely taxonomic information, comparisons of C-banding patterns between species may reveal homeology and other relationships between genomes.

The aim of this study was to investigate C-band variation, both within V. faba and between Vicia species, with view to assessing the origin and evolutionary history of V. faba.

According to Kupicha (1976), Vicia is subdivided into two subgenera, Vicilla (17 sections) and Vicia (sections Atossa, Vicia, Faba, Hypechusa and Peregrinae). Section Faba comprises three groups, distinct both morphologically and karyotypically: i) V. faba, ii) V. bithynica L. and iii) V. narbonensis L. and a complex of species similar to V. narbonensis (V. serratifolia Jacq., V. galilaea Plitm. and Zoh., V. hyaeniscyamus Mouterde and V. johannis Tamamschian). In this study, V. faba was compared to three taxa from section Faba (V. bithynica, V. serratifolia and V. johannis) and also to two species from section Hypechusa (V. lutea L. and V. melanops Sibth. and Sm.). Raina and Rees (1983) demonstrated that V. lutea and V. melanops have amounts of DNA per 2C nucleus more similar to V. faba than are other species from section Faba. These two species may prove more successful than the species in section Faba in interspecific crosses with V. faba (Ramsay and Pickersgill, this volume).

MATERIALS AND METHODS

Seeds were obtained from the sources indicated in Table 1.

Accession		Source
V.faba var.paucijuga	172	Dr.D.A. Bond, PBI, Cambridge, UK
	172	Prof.J. Cubero, ETSIA,Cordoba,Spain
V.faba var.minor	H53/1	Dr.D.A. Bond, PBI, Cambridge, UK
V.faba var.minor	BPL 1192	Dr.M. Saxena, ICARDA, Aleppo,Syria
V.faba var.equina	ILB 337	Dr.M. Saxena, ICARDA, Aleppo,Syria
V.bithynica	855	Dr.C.Lehmann, ZGK,Gatersleben,DDR
V.johannis var.procumbens	64	Dr.C. Lehmann, ZGK,Gatersleben,DDR
V.serratifolia	808289	Dr.F.A. Bisby,Univ. of Southampton,UK
V.lutea	800488	Dr.F.A. Bisby,Univ. of Southampton,UK
V.melanops	800381	Dr.F.A. Bisby,Univ. of Southampton,UK

Table 1. Sources of seed of Vicia species

Roots were collected from young pot-grown plants. C-banding was performed using a method adapted from that of Seal and Bennett (1981). V. faba and V. melanops root tips were pretreated for 3 hours with 0.05% (w/v) colchicine. Other species did not respond adequately to colchicine and were pretreated instead for 3 hours with saturated para-dichlorobenzene. Roots were fixed for 24 hours in 3+1 ethanol + acetic acid and stored, if necessary, in absolute ethanol at 4°C. After rinsing in deionised water, the root tips were hydrolysed in 0.2 M HCl at 20-22°C for 40 minutes. Meristems were then excised in a drop of 45% acetic acid on a slide. Cells were tapped out and squashed under a coverslip. The best slides were selected using phase microscopy. Slides were frozen using the spray Arcton 12 (I.C.I., Runcorn, U.K.) and coverslips were removed with a scalpel and discarded. Slides were quickly placed in absolute ethanol and left for one hour before removing and leaving to dry overnight. The following morning they were treated with saturated Ba(OH)$_2$ at 45°C for 5.5 minutes, rinsed for several minutes with running deionised water, then transferred to a clean dish with 2x SSC (3.506 g NaCl and 1.765 g sodium citrate per 200 ml, adjusted to pH 7.0 with 0.2 M HCl) at room temperature for 15 minutes, after which the solution was replaced with fresh 2x SSC at 52°C for 90 minutes. The slides were then rinsed briefly in stain diluent (1/15 M Na$_2$HPO$_4$) and stained in 2-3% Leishman's stain (30 ml stain mixed with 100 ml diluent, then filtered) for 2-4 hours, after which the slides were rinsed briefly under running deionised water, blotted, air-dried for two hours, warmed on a slide warmer and mounted in Euparal. Leishman's was used in preference to Giemsa because, during trials, bands were obtained more regularly with Leishman's stain.

C-bands were determined for about five good cells for each taxon and their positions plotted on the karyotypes derived from 5-10 root tip cells stained using the Feulgen method. Chromosomes are presented in order of decreasing size and, for V. faba, numbered according to the system proposed by Dobel et al. (1973).

RESULTS

Differences within V. faba

The haploid complement of V. faba contains 6 chromosomes, one metacentric and 5 acrocentrics which cannot be reliably distinguished in conventionally stained preparations.

The C-banded karyotype of V. faba var. equina ILB 337 is presented in Figs. 1a and 2a. The metacentric has two large proximal bands on the short arm which frequently fuse into one, and a small proximal band on the long arm. Major bands on other chromosomes are interstitial. No telomeric bands were detected. This karyotype was also shared by one of the accessions of var. minor (H53/1) but the other accession of var. minor (BPL 1192) was homozygous for a pericentric inversion on the metacentric chromosome (Fig. 3). The single accession of var. paucijuga studied (172) also had a distinguishable metacentric chromosome in which the two bands on the short arm were reduced in intensity (Fig. 3).

Figure 1. C-banded chromosomes of Vicia species.
   a) V. faba var. equina;   b) V. serratifolia;
   c) V johannis; d) V. bithynica;   e) V. lutea
   and f) V. melanops.

Figure 2. C-banded karyotypes of Vicia species.

32

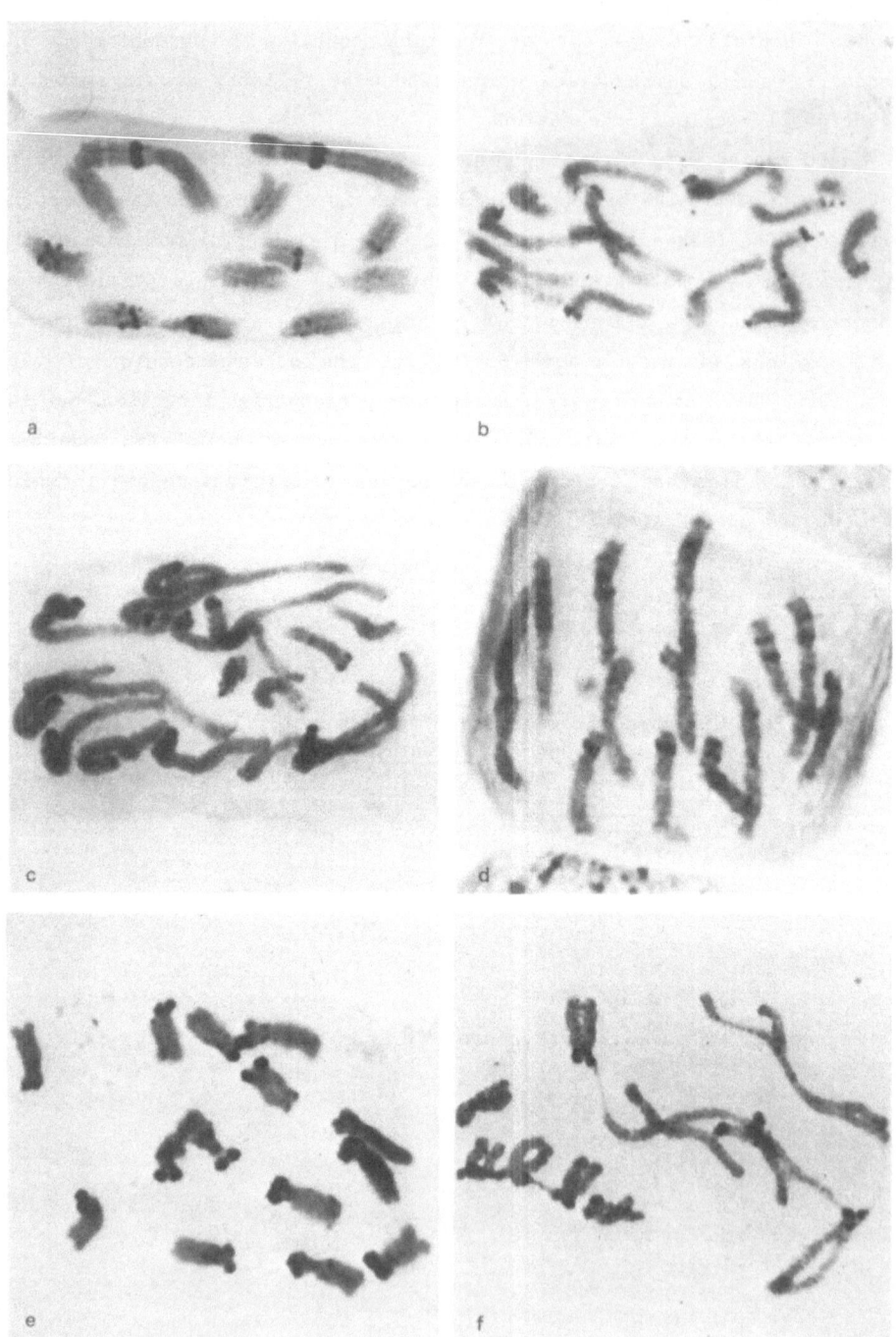

Figure 1

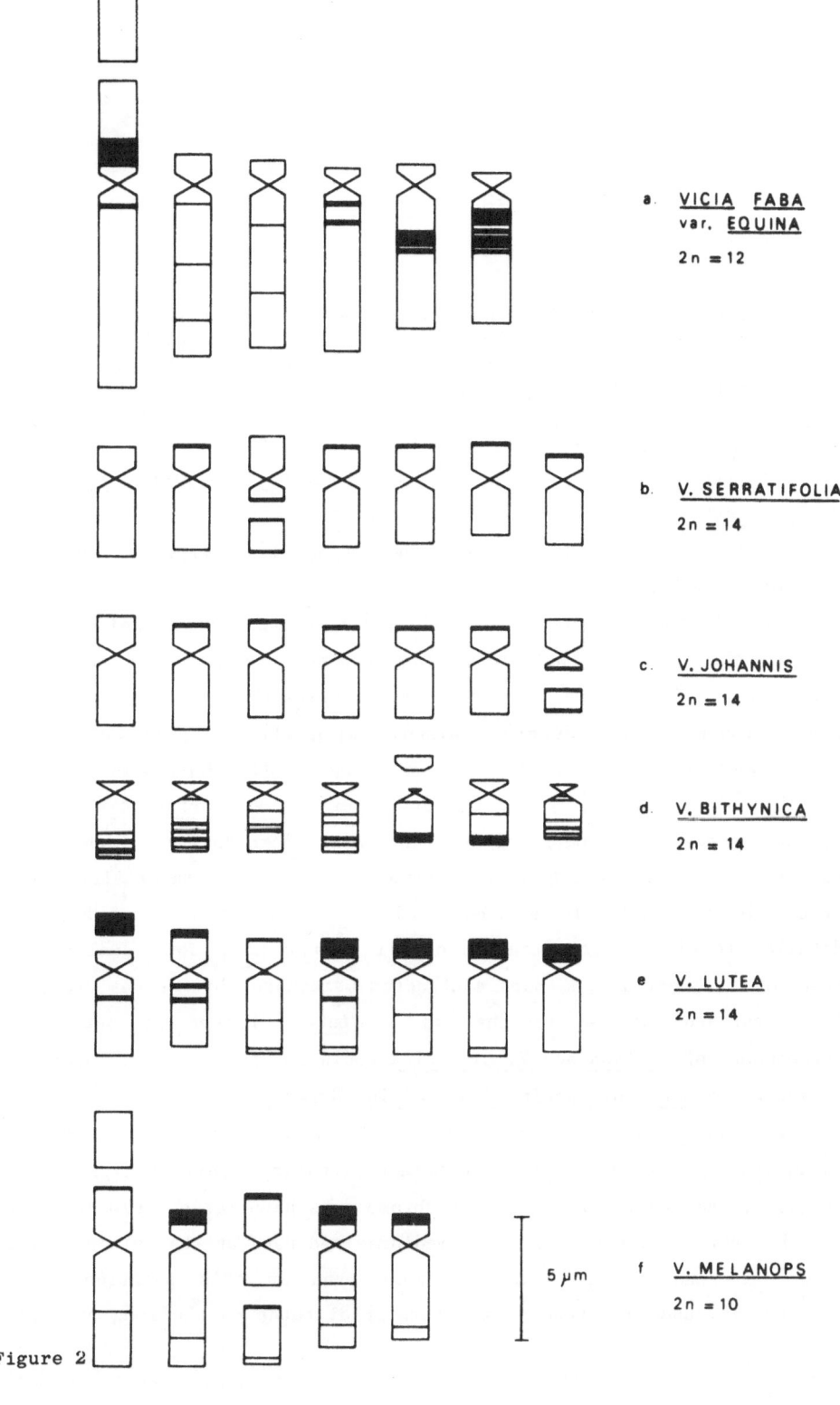

a. **VICIA FABA**
   var. EQUINA
   2n = 12

b. **V. SERRATIFOLIA**
   2n = 14

c. **V. JOHANNIS**
   2n = 14

d. **V. BITHYNICA**
   2n = 14

e. **V. LUTEA**
   2n = 14

5 μm

f. **V. MELANOPS**
   2n = 10

Figure 2

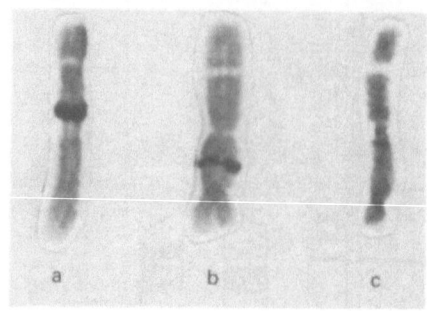

Figure 3.   C-banded metacentric V. faba chromosomes.
a) var. equina ILB 337;   b) var. minor BPL 1192 with
pericentric inversion;   c) var. paucijuga 172 with reduced
intensity of short arm bands.

C-banding patterns of other species from section Faba

The two species studied from the V. narbonensis group, V. johannis
and V. serratifolia (both 2n = 14), have smaller and more submetacentric
chromosomes than those of V. faba. The two species differ from each other
by small changes in karyotype (Figs. 1b, 1c and 2b, 2c), the most notable
of which is the size of the NOR chromosome (nucleolus organiser region-
bearing chromosome). Their C-banding patterns are very similar. All C-
bands are small and are restricted to telomeres of NORs. One telomere of
most chromosomes was deeply stained. With the exception of the NOR
chromosomes, the heterochromatic telomere was on the short arm.

V. bithynica (2n = 14) differs from the V. narbonensis group in
having smaller chromosomes which are acrocentric rather than
submetacentric and which have a complex pattern of interstitial bands
(Figs. 1d and 2d). In prophase cells, the patterns are sufficiently
detailed to allow identification of all chromosomes. Many smaller bands
fuse or disappear in the more contracted metaphase chromosomes, resulting
in a pattern similar to that of V. faba. Homoeology between the
chromosomes of V. faba and V. bithynica could not be detected, however.

C-banding patterns of species from section Hypechusa

V. lutea has seven pairs of similarly-sized submetacentric
chromosomes. They have the most heterochromatin of all species analysed
(Figs. 1e and 2e). Five pairs of chromosomes have large, telomeric bands
in the short arm, one pair has a medium-sized band at the telomere of the
short arm and one pair has a small band in this position. These
differences, and differences in interstitial bands on the long arm, allow

all chromosomes to be identified.

The haploid chromosome set of V. melanops has five chromosomes: a large, metacentric NOR chromosome; a smaller submetacentric NOR chromosome; and three submetacentric to acrocentric chromosomes. These three chromosomes have large heterochromatic telomeres and can be distinguished from each other by a combination of differences in size and interstitial bands. The two larger chromosomes are easily distinguished by size, NOR position and banding patterns (Figs. 1f and 2f).

DISCUSSION

The C-banded karyotype of V. faba presented here is similar to several other published karyotypes (Burger and Sheuermann, 1974; Dobel et al.,1973; Friebe, 1976; Greilhuber, 1975; Hizume et al., 1980; Klasterska and Natarajan, 1975; Pignone and Attolico, 1980; Rowland, 1981; Schweitzer, 1973; Singh and Lelley, 1982; Takehisa and Utsumi, 1973). The major differences between this karyotype and several others are the lack of major bands on chromosomes 2 and 3 and the addition of several minor bands. It is, however, still possible to match the chromosomes with those described by Dobel et al. (1973). The identification of the chromosomes has been confirmed by banding the chromosomes of the reconstructed karyotype 'ACB' of Dobel et al. (1973). It is not clear why these differences in banding pattern occur but it is likely that differences in technique, not plant karyotype, are responsible. This is supported by the lack of the major bands from chromosomes 2 and 3 at the expected sites in preparations of the reconstructed karyotype 'ACB'. These differences may be due to the use of Leishman's, rather than Giemsa, stain and may indicate different staining responses of different classes of heterochromatin.

The results presented here agree, in general, with those of Pignone and Attolico (1980) who also found that the different varieties of V. faba have very similar banding patterns. They reported small differences between varieties in patterns produced by a fluorescent banding technique (quinacrine or Q-banding) which also stains constitutive heterochromatin. A proximal Q-band on chromosome 6 was larger in var. paucijuga than in other types of V. faba. They did not find any differences in C-bands between varieties of V. faba. It has now been shown (Fig. 3) that the two bands on the short arm of the metacentric chromosome are much smaller in this accession of the supposedly primitive var. paucijuga than in other varieties of V. faba. The significance of this is not clear, particularly as the banding patterns of related species are very different. The presence of these polymorphic bands does, however, provide suitable chromosome markers for use in gene mapping.

There are large differences in C-banding patterns between species in section Faba. Perrino and Pignone (1981) found that Q-bands of both V. faba and V. bithynica are all interstitial and on the long arms whereas V. narbonensis, V. serratifolia, V. galilaea and V. johannis all have very small telomeric bands on the short arms. These patterns are very similar to the C-banding patterns reported here. The C-banding data support the suggestion by Yamamoto et al. (1982), based on isozyme studies and acetic-orcein stained karyotype data, that there are three distinct groups within section Faba.

The other two species investigated, V. melanops and V. lutea, are included in section Hypechusa. The C-banding patterns of these species differ from all those in section Faba by having large, terminal C-bands on the short arms of most of their chromosomes. D'Amato et al. (1978) published a C-banded karyotype of V. lutea which shows terminal short arm bands but most of the smaller bands reported here were not observed. Although V. lutea and V. melanops show generally similar banding patterns, there is insufficient similarity in the minor bands to allow the identification of homoeology. The V. melanops karyotype has two chromosomes with long short arms and no large telomeric blocks of heterochromatin. These chromosomes may be the products of the two Robertsonian fusion events which are probably the means by which the haploid chromosome number was reduced to five in this species. C-banded karyotypes of two other species from section Hypechusa, V. hybrida L. and V. pannonica Crantz, were published by D'Amato et al. (1978). These species also had large, telomeric bands on the short arms which indicates that banding patterns may be a sectional characteristic.

The relationship between V. faba and other species of Vicia remains obscure. No homoeology can be detected between the chromosomes of V. faba and other species of section Faba using C-banding patterns. In addition to changes in the positions of the major blocks of heterochromatin, the chromosome arm ratios differ greatly between V. faba and related species of the V. narbonensis group.

There are two alternative ways in which the V. faba karyotype could have evolved from a putatively primitive one (Fig. 4). A primitive Vicia karyotype would have the ancestral basic number for the genus and would be likely to resemble the majority of extant Vicia species in banding pattern and chromosome morphology (i.e. n = 7, terminal heterochromatin and submetacentric chromosomes). One possible way of obtaining a karyotype similar to V. faba is by the incorporation of sub-telomeric heterochromatin into the long arms of the chromosomes by pericentric inversions (Fig. 4a). Chromosome pairing in hybrids between V. faba and species of the V. narbonensis group, should they become available, may confirm or refute this suggestion.

A   Inversion

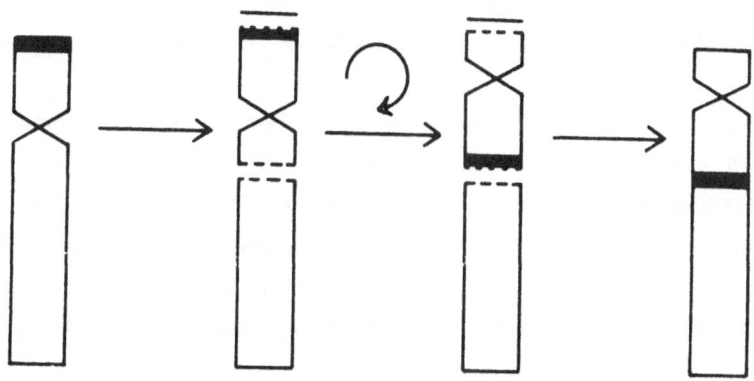

B   Equilocal heterochromatin

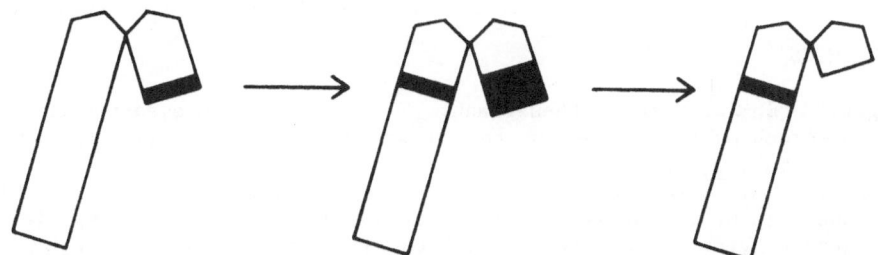

Figure 4. Possible derivation of V. faba karyotype

An alternative theory depends on the observations of Heitz (1932) and Greilhuber et al. (1981) that heterochromatin tends to be equilocal from the centromere in distribution. According to Greilhuber et al. (1981), heterochromatin accumulates as some species evolve and new sites are formed on different chromosomes or on the other arm of the same chromosome at the same distance from the centromere as pre-existing heterochromatin blocks. If this occurred in a species possessing telomeric heterochromatin on the short arm, as in many species of Vicia, and the heterochromatic telomeres were subsequently deleted, a karyotype similar to V. faba or V. bithynica could arise (Fig. 4b). This scheme assumes that heterochromatin can both accumulate and be deleted in the course of evolution. Spontaneous deletion of heterochromatin telomeres has occurred, for example, in several lines of triticale (Seal and Bennett, 1981). In addition, Lavania and Sharma (1983) give several examples of the reduction of heterochromatin with phylogenetic advancement, so this scheme for the derivation of a karyotype similar to that of V. faba seems plausible.

C-banding has been demonstrated to be a very useful technique in genetical and evolutionary studies in Vicia, and is already known to be useful in several other genera. The technique remains one of the best ways of identifying chromosomes or chromosome segments of species with suitable banding patterns, such as V. faba, and is likely to be extremely useful in the future as an aid to a variety of cytogenetical studies.

ACKNOWLEDGEMENTS

The work reported here forms part of a project on interspecific hybridisation involving V. faba for which financial support from the Overseas Development Administration is gratefully acknowledged. I would also like to thank Dr. B. Pickersgill for her comments on the manuscript.

REFERENCES

Burger, E.-C., and Sheuermann, W. 1974. Giemsa-Banden und heterochromatische Region bei Metaphasechromosomen von Vicia faba. Cytobiol. 9, 23-35.

D'Amato, G., Bianchi, G., Caperini, R. and Marchi, P. 1978. Localizzazione delle bande C nel cariotipo di cinque specie del genere Vicia e due del genere Lathyrus. Annali di Botanica 37, 189-201.

Dobel, P., Rieger, R. and Michaelis, A. 1973. The Giemsa banding patterns of the standard and four reconstructed karyotypes of Vicia faba. Chromosoma 43, 409-422.

Friebe, B. 1976. Specifische Giemsa-Farbung von heterochromatischen Chromosomensegmenten bei Vicia faba, Allium cepa und Paeonia tenuifolia. Theor. Appl. Genet. 47, 275-283.

Greilhuber, J. 1975. Heterogeneity of heterochromatin in plants: Comparison of Hy- and C-bands in Vicia faba. Pl. Syst. Evol. 124 139-156.

Greilhuber, J., Deumling, B. and Speta, F. 1981. Evolutionary aspects of chromosome banding, heterochromatin, satellite DNA and genome size in Scilla (Liliaceae). Ber. Deutsch. Bot. Ges. Bd. 94, 249-266.

Heitz, E. 1932. Die Herkunft der Chromozentren. Planta 18, 571-636.

Hizume, M., Tanaka, A., Yonezawa, Y., Tanaka, R. 1980. A technique for C-banding in Vicia faba chromosomes. Jap. J. Genet. 55, 301-305.

Klasterska, I. and Natarajan, A.T. 1975. Distribution of heterochromatin in the chromosomes of Nigella damascena and Vicia faba. Hereditas 79, 154-156.

Kupicha, F.K. 1976. The infrageneric structure of Vcia. Notes Royuy. Bot. Gard. Edinb. 34, 287-326.

Lavania, U.C. and Sharma, A.K. 1983. Chromosome banding in evolutionary plant cytogenetics. Proc. Indian Acad. Sci. (Plant Sci.) 92, 51-79.

Linde-Laursen, I. 1979. Giemsa C-banding of barley chromosomes. III. Segregation and linkage of C-bands on chromosomes 3, 6 and 7. Hereditas 91, 73-77.

Perrino, P. and Puignone, D. 1981. Contribution to the taxonomy of Vicia species belonging to the section Faba. Die Kulturpflanze 29, 311-319.

Pignone, D. and Attolico, M. 1980. Chromosome banding in four groups of Vicia faba L. Caryologia 33, 283-288.

Raina, S.N. and Rees, H. 1983. DNA variation between and within chromosome complements of Vicia species. Heredity 51, 335-346.

Ramsay, G. and Pickersgill, B. (this volume). Interspecific hybridisation in Vicia section Faba: Comparison of selfed and hybrid embryo and endosperm.

Rowland, R.E. 1981. Chromosome banding and heterochromatin in V. faba. Theor. Appl.Genetic. 60, 275-280.

Schweitzer, D. 1973. Differential staining of plant chromosomes with Giemsa. Chromosoma 40, 307-320.

Seal, A.G. and Bennett, M.D. 1981. The rye genome in winter hexaploid triticales. Can. J. Genet. Cytol. 23, 647-653.

Singh, V.P. and Lelley, T. 1982. Giemsa C-banding karyotype of V. narbonensis as compared to V. faba. FABIS 4, 22-23.

Takehisa, S. and Utsumi, S. 1973. Visualisation of metaphase heterochromatin in Vicia faba by the denaturation-renaturation Giemsa staining method. Experientia 29, 120-121.

Yamamoto, K., Moritani, O. and Ando, A. 1982. Karyotypic and isozymatic polymorphism in species of the section Faba (genus· Vicia). Tech. Bull. Fac. Agric. Kagawa Univ. 34, 1-12.

PRETREATMENT OF <u>VICIA FABA</u> ROOT TIP MERISTEMS WITH LOW CLASTOGEN DOSES PROTECTS AGAINST ABERRATION INDUCTION BY SUBSEQUENT TREATMENTS: INDUCTION OF REPAIR PROCESSES?

R. Rieger[1], A. Michaelis[1], H. Nicoloff[2]

[1]Zentralinstitut fur Genetik und Kulturpflanzenforschung der Akademie der Wissenschaften der DDR, DDR-4325 Gatersleben, German Democratic Republic

[2]Institute of Genetics, Bulgarian Academy of Sciences, 1113, Sofia, Bulgaria.

ABSTRACT

In order to counteract potential DNA damage, organisms have evolved mechanisms to reverse, remove or tolerate such adverse influences. A distinction is made between 'constitutive' mechanisms, active at all times and 'inducible' ones that represent response to damage. The authors explore in this paper the extent to which low doses of clastogens can protect the system against higher doses. They show that one alkylating agent may substitute for another but that conditioning to induce protective effects is prevented by inhibition of protein synthesis.

INTRODUCTION

Root tip meristems of <u>Vicia faba</u> are extensively being used to investigate, among others, the mode and specificity of action of clastogens inducing chromosome structural changes (cf. Kihlman, 1982; Schubert <u>et al</u>., 1982) in plants. In previous experiments (Rieger <u>et al</u>., 1982), we tested the influence of pretreatment with low doses of certain clastogens ("conditioning" treatment) on the yield of chromatid aberrations induced by a higher dose ("challenge" treatment) of the same or of different clastogens applicated to <u>Vicia faba</u> main roots two hours later. As compared to the controls ("conditioning" treatment and "challenge" treatment alone, respectively) the main results obtained may be summarized as follows:

1. "Conditioning" pretreatment of <u>Vicia faba</u> root tip meristems may exert protective effects, i.e. the yield of aberrations produced by the consecutive "challenge" treatment may become significantly reduced by "conditioning". Only a very low frequency of cells with chromatid aberrations was induced by the "conditioning" treatment (the

spontaneous frequency of metaphases with aberrations in V. faba root tip meristems is about 1%).

2. Availability of "below-additivity effects" with respect to the yield of induced chromatid structural changes after "conditioning" and "challenge" treatments with N-methyl-N-nitrosourea (MNU), ethylmethane sulfonate (EMS), triethylenemelamine (TEM), maleic hydrazide (MH) and ethanol (EA) depends on the clastogen combination being used. No protective effects were observed when EA was used for "conditioning" and "challenge" treatments. All other clastogens tested resulted in protective effects, i.e. "conditioning" treatments reduced the yield of chromatid aberrations induced by "challenge" treatment.

3. All alkylating agents tested were able to substitute for each other in inducing protection. They were unable to do so when MH ws used for "challenge" treatment and vice versa, although "conditioning" with MH resulted in a significant reduction of the aberration yield induced by MH "challenge" treatment. So whatever gives rise to protection by "conditioning", the effect is clearly clastogen-dependent: some clastogens can substitute for each other in provoking the effect, others cannot.

With the help of the experiments to be reported here, we tried to find answers on some questions connected with the protective effects of pretreatments with low doses of clastogens:
-   Is the effect dependent on the clastogen concentration used for "conditioning"?
-   Does effective "conditioning" with a certain clastogen concentration decay with prolongation of the time span between "conditioning" and "challenge" treatment?
-   Can protection by "conditioning" be influenced by inhibition of protein synthesis in the root tip meristems?
Answers to the first two questions were derived from treatment with MH, those on the third question from treatments with TEM and cycloheximide. All three questions found positive answers and suggest a mechanism of protection via "repair" processes induced by "conditioning" treatment.

MATERIALS AND METHODS

Primary root tip meristems of the reconstructed karyotype ACB (cf. Michaelis & Rieger, 1971) were used in the present experiments and treated with maleic hydrazide (1,2-dihydro-pyridazine-3,6-dione) or TEM (triethylenemelamine) for "conditioning" and "challenge" at 24° C (treatment times, concentrations of MH, TEM and cycloheximide, as well as intervals between treatments as stated in the legends). After treatment and various recovery times (15, 18, 21, and 24 h) in running tap water (24° C), the roots were treated with 0.05% colchicine for two hours. Root tips (length about 2-3 mm) were fixed in 3 parts absolute ethanol and 1 part glacial acetic acid. Feulgen squashes were made permanent by the dry-ice method.

For each recovery time (RT) the percentage of metaphases (all from the first cell cycle after treatment) with chromatid aberrations was established and the following aberration types scored: isochromatid breaks, chromatid translocations, duplication deletions, and intercalary deletions. 50 metaphases from each of four root tips (= 200 cells) and recovery time were checked for the presence of chromatid aberrations; at least two repetitions of each experiment were made and the results pooled.

RESULTS

Maleic hydrazide is a growth inhibitor and clastogen in plants. It is used in agriculture as a herbicide and plant growth regulator (cf. Schone and Hoffmann, 1949). Its clastogenic activity was originally reported by Darlington and McLeish (1951) in Vicia faba and has since then been confirmed for all other plant species investigated (cf. Swietlinska and Zuk, 1978). MH is an S-phase dependent clastogen inducing chromatid structural changes preferentially involving specific chromosome regions (cf. Heindorff and Rieger, 1984). Its mode of action eventually resulting in chromatid aberrations is unknown (cf. Heindorff and Rieger, 1984) but clearly dependent on the metabolic state of the cells treated (Kihlman, 1956; 1966).

1. Dependence of protection on the MH concentration used for "conditioning" treatment

Fig. 1 summarises the evidence for the dependence of protection on the MH concentration used for "conditioning". When "conditioning" and "challenge" (0.5 h, 4 x 10⁻⁴ M) treatments were separated by two hours and pretreatment (0.5 h) was done with MH concentrations of 5 x

10^{-5} M, 10^{-5} M, 10^{-6} M and 10^{-9} M, the last-mentioned concentration proved to be ineffective in provoking protective effects. The other "conditioning" concentrations of MH unambiguously resulted in protection against the "challenge" treatment, as measured by the reduced yield of metaphases with induced chromatid structural changes.

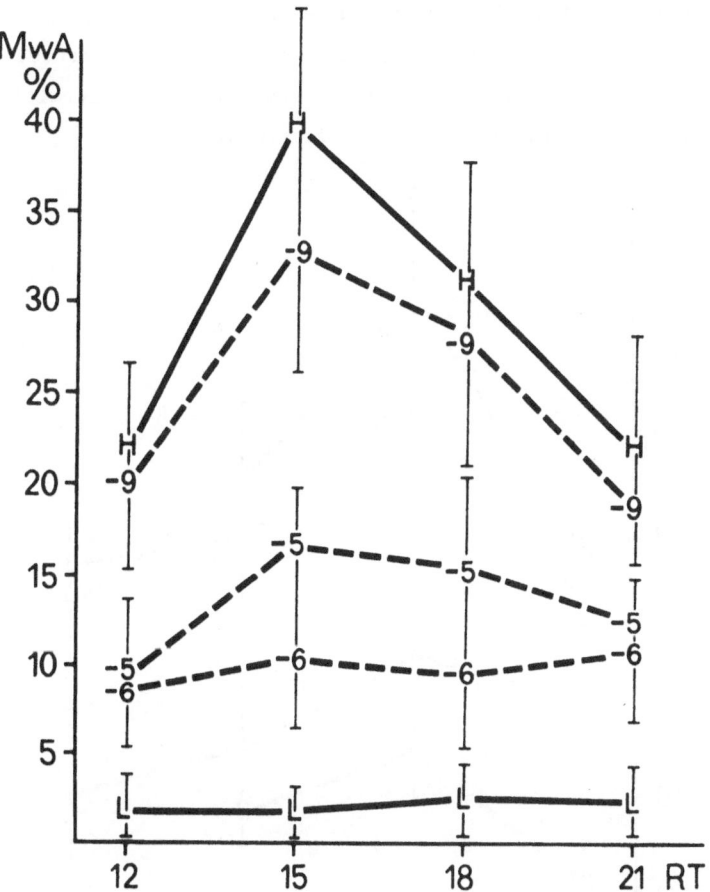

Figure 1. The frequency (%) of metaphases with chromatid aberrations after "conditioning" (0.5h, 5 x 10^{-5}M) treatment with MH (curve symbolized by L), "challenge" (0.5h, 4 x 10^{-4} M) treatment with MH (curve symbolized H), and consecutive treatments with the low and high MH concentrated separated 2h (broken lines: first treatment with 10^{-5}M, 10^{-6}M, and 10^{-9}M MH, respectively, second treatment with 4 x 10^{-4}M MH). RT: recovery time after second treatment in hours.

44

2. Dependence of protection on the time span between "conditioning" and "challenge" treatment

Induction of protection by "conditioning" treatment with MH proved to be dependent on the time span between "conditioning" and "challenge" treatment (Fig. 2). The MH concentration used for "conditioning" was 5×10^{-5} M (0.5 h), the "challenge" concentration 4×10^{-4} M (0.5 h). The time interval between the two treatments ranged from 2 to 8 hours (2, 4, 8 h). Protective effects of "conditioning" are evident for time spans of 2 and 4 h between the two treatments. When the treatments were separated by 8 h, protection by "conditioning" was absent, i.e. the effect obviously decays with time.

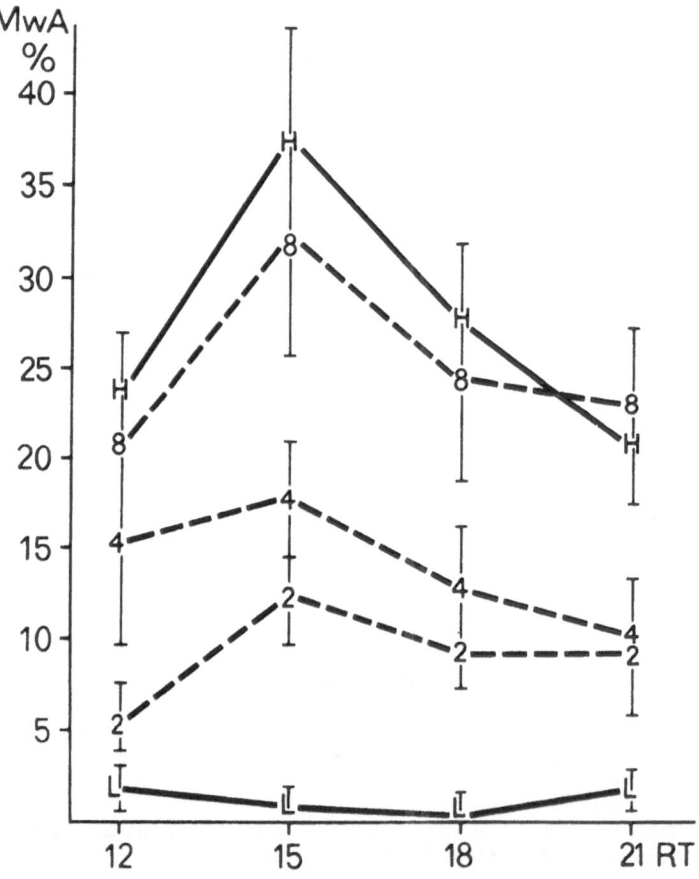

Figure 2. The influence of various time intervals (2, 4 and 8h) between "conditioning" (0.5h, 5×10^{-5}M MH) and "challenge" (0.5h, 4×10^{-4}M MH) treatment on the frequency of chromatid aberrations observed (L = "conditioning" treatment alone; H = "challenge" treatment alone, broken lines: aberration yields for time intervals of 2, 4 and 8h).

This conclusion is substantiated by another set of experiments (Fig. 3) in which "conditioning" was done with 10^{-5} M MH (0.5 h), "challenge" with 4×10^{-4} M MH (0.5 h) and the time interval between the two treatments was 2 or 4 hours. Under these conditions, the reduced concentration of MH (10^{-5} M instead of 5×10^{-5} M) resulted in disappearance of protective effects already with a 4 h time span between "conditioning" and "challenge" treatment (with 5×10^{-5} M MH, protection disappeared only when the interval between the treatments was 8 h).

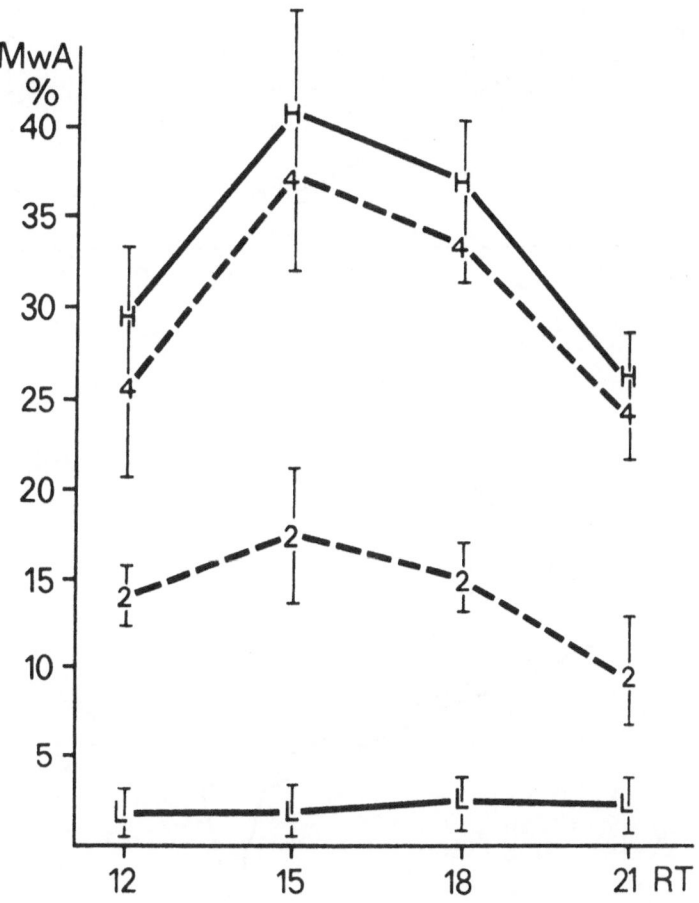

Figure 3. The influence of a prolonged time interval (from 2 to 4h) between "conditioning" (0.5h, 10^{-5}M MH) and "challenge" (0.5h, 4×10^{-4}M MH) on the frequency of induced chromatid aberrations (L = "conditioning" treatment alone; H = "challenge" treatment alone; broken lines: consecutive treatments separated by 2 and 4h, respectively).

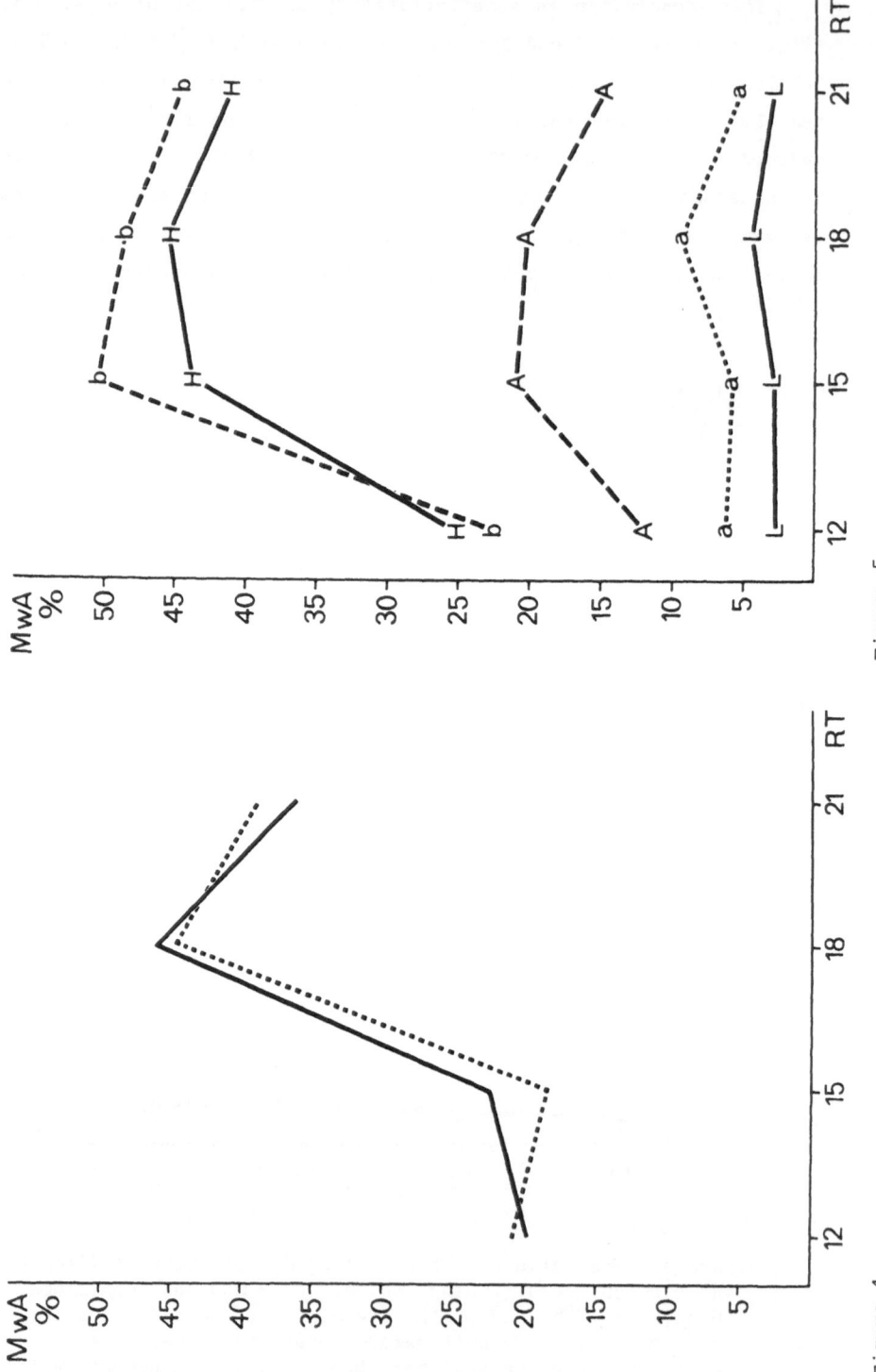

Figure 5

Figure 4

Figure 4. Percent metaphases with chromatid aberrations after treatment with TEM (0.5h, 10^{-4}M; solid line) and with pretreatment (2h) by cycloheximide (10^{-6}M) prior to TEM treatment (0.5h, 10^{-4}M; broken line). RT: recovery time in hours

Figure 5. Protective effects (curve A) after TEM "conditioning" (curve L = "conditioning" treatment with 10^{-5} M TEM for 0.5h; curve H = "challenge" treatment with 10^{-4} M TEM for 0.5h) as affected by pretreatment (2h) with 10^{-6}M cycloheximide (curve b) prior to "conditioning". Cycloheximide pretreatment completely abolished protective effects exerted by "conditioning" pretreatment. Cycloheximide treatment (2h,10^{-6}M; curve a) is weakly clastogenic itself. (RT: recovery time ; MwA : percent metaphases with chromatid aberrations).

3. Effects of cycloheximide on protection induced by "conditioning" with TEM

TEM is a polyfunctional alkylating agent whose clastogenic activity is expected to remain relatively uninfluenced by inhibitors of protein synthesis (cf. Rieger and Michaelis, 1967). This is borne out by the fact (Fig. 4) that aberration induction by TEM is quantitatively the same in the presence and absence of cycloheximide pretreatment (2 h, 10^{-6} M cycloheximide). When "conditioning" (0.5 h, 10^{-5} M TEM) and "challenge" (0.5 h, 10^{-4} M TEM) treatments were separated by two hours, "conditioning" treatment with TEM protected against the TEM "challenge" treatment as evident from the drastically reduced frequency of metaphases with chromatid aberrations (Fig. 5) scored under these circumstances. Cycloheximide alone proved to be weakly clastogenic (2, 10^{-6} M) in Vicia faba and applied in this concentration prior to "conditioning", completely prevented protective effects exerted by "conditioning" in the absence of cycloheximide; lower cycloheximide concentrations (10^{-7} M) proved to be unable to do so (data not shown). These results indicate that unimpaired protein synthesis is a prerequisite for the induction of protective effects by "conditioning" TEM treatment.

Similar observations have been made in mammalian cells cultivated in vitro: cycloheximide abolished the protective effects of the "conditioning" pretreatments with various methylating agents (Kaina, 1983).

DISCUSSION AND CONCLUSIONS

In order to counteract potentially damaging effects, a series of mechanisms has evolved, by which cellular systems and cells can reverse, remove or tolerate such adverse influences. These mechanisms are of fundamental importance for the protection of pro- and eukaryotes from genetic alterations and for survival of cells, organisms, and species. Several repair mechanisms of DNA damage (cf. Rieger et al., 1982) are well characterized; some are active in the cell at all times (constitutive mechanisms), others are inducible when the cell incurs certain kinds of damages (inducible mechanisms).

Though the underlying mechanisms are presently unknown, the protective effects exerted by a low "conditioning" dose of certain clastogens on the clastogenic effectivity of a higher "challenge" dose suggest the involvement of inducible repair systems. The experimental results described here allow the following conclusions:

- Induction of protective effects by "conditioning" treatment is dependent on the clastogen concentration being used; below a certain threshold no such effect occurs.
- Protective effects decay with the length of the time span between "conditioning" and "challenge" treatment; the length of this interval is dependent on the clastogen dose being used for "conditioning".
- The induction of protective effects by "conditioning" is prevented, at least in the case of TEM, by inhibition of protein synthesis.

Taken together, these observations are indicative of protective processes which become induced by "conditioning" treatments with low doses of clastogens which per se do not result in significantly increased yields of chromatid structural changes. The underlying mechanisms are presently completely unknown but protein synthesis seems to be involved in the induction of protective effects. The data reported here and earlier

(Rieger et al., 1982) are the first with respect to protection by clastogen pretreatment against the chemical induction of chromatid structural changes in plants, but similar observations, derived from the analysis of other biological end effects studied after treatment with ionizing radiation (Sybenga and Kleijer, 1976; Broertjes, 1972; Leenhouts et al., 1982) and chemical mutagens (Veleminsky et al., 1983) are at hand. They suggest that protection induced by "conditioning" treatment may be a general phenomenon taking influence on cell survival, induction of gene mutations and chromosome structural changes in prokaryotes and eukaryotes.

The presently best characterized inducible activity in bacteria is called "adaptive response" (Jeggo et al., 1977) and can effectively reduce mutagenesis. The system is able to remove both ethyl and methyl group from O^6-alkylguanine (a^6G) in DNA and makes the cells less sensitive to high "chalenge" doses of simple alkylating agents (Samson and Cairns, 1977; Schendel and Robins, 1979; Sedgwick, 1983). The adaptive response of bacteria is due to the synthesis of an acceptor protein forming S-alkylcysteine at an acceptor site on the same protein that catalyzes the reaction (Lindahl et al., 1982; Mitra et al., 1982). Basically similar observations have been made in mammalian cells cultivated in vitro (cf. Waldstein et al., 1982; Kaina, 1983).

The data at hand for plant cells show that the formally analogous effects of "conditioning" pretreatments are not confined to the use of alkylating agents and that different agents cannot in all cases substitute for each other in provoking protection. Thus, different agents used for "conditioning" may in fact induce different systems eventually resulting in protection against "challenge" treatments and for that reason cannot substitute for each other. Whether or not induced protection is due to inducible DNA repair processes is still an open question which deserves further study.

REFERENCES

Broertjes, C. 1972. Use in plant breeding of acute, chronic or fractionated doses of X-rys or fst neutrons as illustrated with leaves of Saintpaulia. Agric. Res. Rep. 776, pp. 74, Centre for Agricultural Publishing and Documentation.
Darlington, C.D., McLeish, I. 1951. Action of maleic hydrazide on the cell. Nature (Lond.), 167, 407-408.

Heindorff, K., Rieger, R. 1984. Exogenous factors affecting yield and intrachromosomal distribution of maleic hydrazide-induced chromatid aberrations in Vicia faba. Biol. Zbl. 103, 9–23.

Jeggo, P., Defais, M., Samson, L., Schendel, P. 1977. An adaptive response of E. coli to low levels of alkylating agents; comparison with previously characterized DNA repair pathways. Molec. Gen. Genetics 157, 1–9.

Kaina, B. 1983a. Cross-resistance studies with V79 Chinese hamster cells adapted to the mutagenic or clastogenic effect of N-methyl-N'-nitro-N-nitrosoguanidine. Mutat. Res. 111, 341–352.

Kaina, B. 1983b. Studies on adaptation of V79 Chinese hamster cells to low doses of methylating agents. Carcinogenesis 4, 1437–1443.

Kihlman, B.A. 1956. Factors affecting the production of chromosome aberrations by chemicals. J. Biophys. Biochem. Cytol. 2, 543–555.

Kihlman, B.A. 1966. Actions of Chemicals on Dividing Cells. Prentice-Hall, Inc., Englewood Cliffs, New Jersey, pp. 260.

Kihlman, B.A. 1982. Root tips of Vicia faba as a material for studying the induction of chromosomal aberrations and sister chromatid exchanges. In : Hsu, T.C. (ed.) 'Cytogenetic Assays of Environmental Mutagens'. Allanheld, Osmun Publishers, 81–100.

Leenhouts, H.P., Broertjes, C., Sijsma, M.J., Chadwick, K.H. 1982. Radiation stimulated repair in Saintpaulia: its cellular basis and effect on mutation frequency. Environmental and Experimental Bot. 22, 301–306.

Lindahl, T., Rydberg, B., Hjelmgren, T., Olsson, M., Jacobson, A. 1982. Cellular defense mechanisms against alkylation of DNA. In: Lemontt, J. and Generoso, W.M. (eds.) Molecular and cellular Mechanisms of Carcinogenesis. Plenum Press, New York, 89–103.

Michaelis, A., Rieger, R. 1971. New karyotypes of Vicia faba L. Chromosoma 35, 1–8.

Mitra, S., Pal, B.C., Foote, R.S. 1982. O^6-methylguanine-DNA methyltransferase in wild-type and ada mutants of E. coli Journ. Bacteriol. 152, 534–537.

Rieger R., Michaelis, A. 1967. Die Chromosomenmutationen. In: Genetik – Grandlagen, Ergebnisse und Probleme in Einzeldarstellungen (Hrsgb. H. Stubbe), VEB Gustav Fischer Verlag, Jena.

Rieger, R., Michaelis, A., Nicoloff, H. 1982. Inducible repair processes in plant root tip meristems? "Below-additivity effects" of unequally gractionated clastogen concentration. Biol. Zbl. 101, 125–138.

Samson, L., Cairns, J. 1977. A new pathway for DNA repair in E. coli. Nature (Lond.) 267, 281–283.

Schendel, P.F., Robins, P.E. 1978. Repair of O^6-methylguanine in adapted E. coli. Proc. Natl. Acad. Sci. USA 75, 6017–6020.

Schone, D.L., Hoffman, O.L. 1949. Maleic hydrazide, a unique growth regulant. Science 109, 588–590.

Schubert, I., Rieger, R., Michaelis, A. 1982. Reconstructed karyotypes as a tool for investigating differential chromosome mutagen sensitivity. In: Hsu, T.S. (ed.) 'Cytogenetic Assays of Environmental Mutagens', Allanheld, Osmun Publishers, pp. 101–105.

Sedgewick, B. 1983. Molecular cloning of a gene which regulates the adaptive response to alkylating agents in E. coli. Molec. Gen. Genetics 191, 466–472.

Swietlinsk, Z., Zuk, J. 1978. Cytotoxic effects of maleic hydrazide. Mutation Res. 55, 15–30.

Sybenga, J., Kleijer, G. 1976. "Below-additivity" and "protective" effects of dose fractionation in Crotolaria intermedia. Mutation Res. 34, 131–140.

Veleminsky, J., Gichner, T., Satava, J. 1983. Reduction in the frequency of N-methyl-N-nitrosoure-induced somatic mutations in _Tradescantia_ by pretreatment with low doses of alkylating agents. Mutation Res. 122, 229-234.

Waldstein, E.A., Cao, E.H., Setlow, R.B. 1982. Adaptive increase of O^6-methylguanine-acceptor proteins in HeLa cells following N-methyl-N'-nitro-N-nitrosoguanidine treatment. Nucleic Acid Res. 10, 4595-4605.

KARYOTYPIC ANALYSIS OF A MUTANT OF <u>VICIA FABA</u> L. BY MICRODENSITOMETRIC PROCEDURE

E. Filippone*, C. Conicella*, A. Errico**, F. Saccardo***

*Centro Miglioramento Genetico Ortaggi C.N.R. Portici (Napoli)

**Istituto di Agronomia, Miglioramento Genetico,
Universita di Napoli

***E.N.E.A., Casaccia, Roma

ABSTRACT

Cytological analysis was carried out on an oblong-leaflet mutant obtained in <u>Vicia faba</u> in the progeny of an M_1 variegated plant characterized by a dicentric chromosome. The karyotype of the mutant was derived by metric and photometric data through a microdensitometer scanner: a longer centromeric region of chromosome I was found in the mutant. The origin of the mutation is discussed.

INTRODUCTION

Several X- and gamma-irradiations of mature pollen grains were performed in <u>Vicia faba</u> (Scuro Torre Lama ecotype) in order to induce new variability.

Morphological and chromosome mutations were induced and isolated in the different generations. Particular care was given to a variegated M_1 plant obtained with 750 Rad of X-rays and to its progeny, following a methodology already developed for peas (Monti et al., 1969; Saccardo et al., 1974; Monti et al., in press).

This paper reports the morphological data and the cytological analysis of a mutant with a modified leaf shape carried out by the use of a microdensitometer apparatus.

MATERIALS AND METHODS

The cytological and phenotypic analyses were performed on M_2, M_3 and M_5 mutated plants all of them characterized by oblong-leaflets.

Meiotic analysis was carried out on the M_2 plant and on four M_3 plants one of which was variegated: squashes of anthers were fixed for 48 hours in ethanol:propionic acid 3:1, to which a small amount of ferric chloride had been added as a mordant and stained with acetocarmine.

Pollen fertility of the same plants was also estimated by staining with acetocarmine.

Mitotic analysis was done on the same M_2 and M_3 plants on squashes of 1.5-2 cm long root tips pretreated with 0.05% solution of colchicine for 7 hours at 3°C, fixed in ethanol-acetic acid 3:1 for at least 2 hours and stained with the Feulgen technique (hydrolysis in 1NHC1 at 60°C for 7 min).

Mitotic metaphase chromosomes were photographed using an "Agfa Ortho 25" film. Six photographic plates were analyzed as described previously (Filippone et al., 1983) by a digital microdensitometer, Perkin & Elmer PDS 1010/A. Densitometric data were subsequently processed by a minicomputer Digital PDP 11/34. For each chromosome, metric (total length and arm ratio) and photometric parameters were obtained and recorded on magnetic tape; the relative length was determined as the ratio between each chromosome and the chromosome 6 which is the shortest one.

M_5 seedlings derived from the M_3 plant No. 1 which did not show any cytologic irregularity were used for karyotype analysis, together with control plants.

	CONTROL		MUTANT	
Plants analyzed (no.)	10		6	
Leaflet length (cm)	7.05	1.83	9.86*	1.12
Leaflet width (cm)	3.93	2.41	3.67	1.50
Leaflet width/length ratio	0.54	0.07	0.41	0.03
Internode length (cm)	6.5	0.7	7.8	1.0
Plant height (cm)	174	16	188	9

*Significantly different from control at the 1% level

Table 1   Main traits of the control and of the oblong-leaf mutant.

RESULTS

Table 1 reports the main traits of the oblong-leaflet mutant (Fig. 1a) in comparison with the control plants: significant differences were only found for the length of the leaflets, which were longer in the mutant (9.9 cm) than in the control (7.0 cm) leaves.

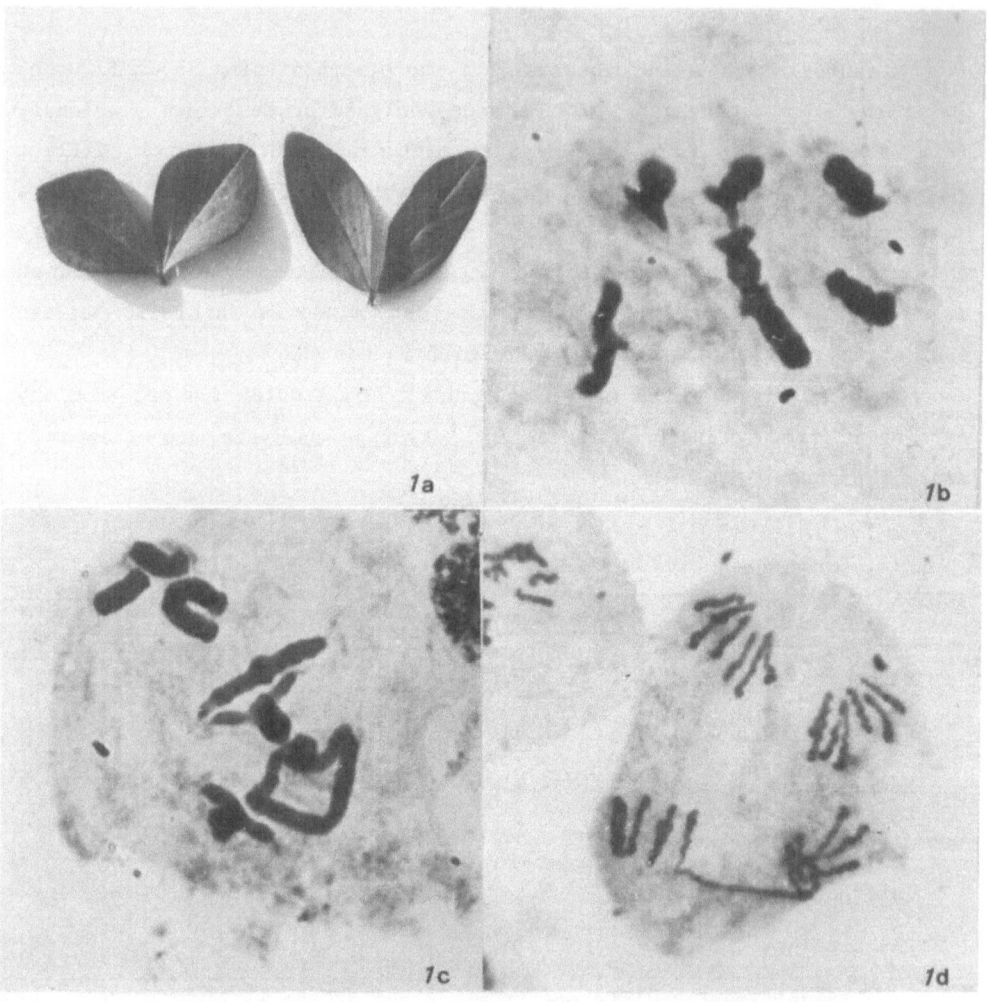

Figure 1. a) Leaflets of the control (left) and of the mutant.
 b) Diakinesis with 6 bivalents in the M$_3$ plant
       no. 1.
 c) Metaphase I with 5 bivalents plus 2 univalents in
       the M$_3$ plant no. 3.
 d) Anaphase II with a simple bridge.

Meiotic analysis of the mutant plants is reported in Table 2: normal diakinesis or metaphase 1 were found in all the plants (Fig. 1b) except plant $M_3 - 3$ which showed univalents (11%) (Fig. 1c); in anaphase II, three plants out of four showed simple bridges (Fig. 1d) and fragments, while plant $M_3 - 1$ was normal. A reduced pollen fertility of about 50% was presented by the $M_3 - 3$ plant which showed also a variegated leaf phenotype.

Plant	DIAKINESIS			ANAPHASE II				
	Cells analyzed no.	Cells with 611 %	511+21 %	Cells analyzed No.	Normal cells %	Cells with bridges %	fragments %	Pollen fertility %
Control	45	100	-	38	100	-	-	98
M_2 1	50	100	-	81	94	3	3	90
M_3 1	77	100	-	41	100	-	-	98
2	10	100	-	14	71	21	8	93
3	34	89	11	30	87	6	7	56
4	34	100	-	13	69	16	15	98

Table 2. Meiotic analysis of the oblong-leaflet mutant in M_2 and M_3 generations

The analysis of the metaphase chromosomes of the control and of the mutant plants shows (Table 3) a statistically significant difference in the length of chromosome 1, that of the mutant being longer by about 9%; no other differences were found in the lengths of the other chromosomes and for the arm ratios.

The densitometric profiles of the centromeric region of chromosome 1 is shown in Fig. 2 for both the mutant and the control: a longer centromeric region of about 6% was found in the mutant.

Chrom. No.	Length		Arm-ratio	
	Control	Mutant	Control	Mutant
1	235.2 ± 9.3	259.8* ± 13.5	1.6 ± 0.2	1.5 ± 0.1
2	123.3 ± 3.0	127.7 ± 4.5	5.3 ± 1.2	5.2 ± 1.2
3	118.3 ± 2.7	120.4 ± 7.3	5.8 ± 1.7	4.9 ± 0.7
4	112.5 ± 3.8	113.0 ± 3.9	6.0 ± 1.7	5.6 ± 1.2
5	105.6 ± 2.4	108.3 ± 2.7	6.1 ± 1.5	6.2 ± 0.6
6	100.0	100.0	5.4 ± 1.1	4.9 ± 0.6

* Significantly different from control
at 5% level

Table 3. Length (as % of chromosome 6) and arm-ratio of the
chromosomes in the control and in the oblong-leaflet
mutant. Data from six processed metaphase plates.

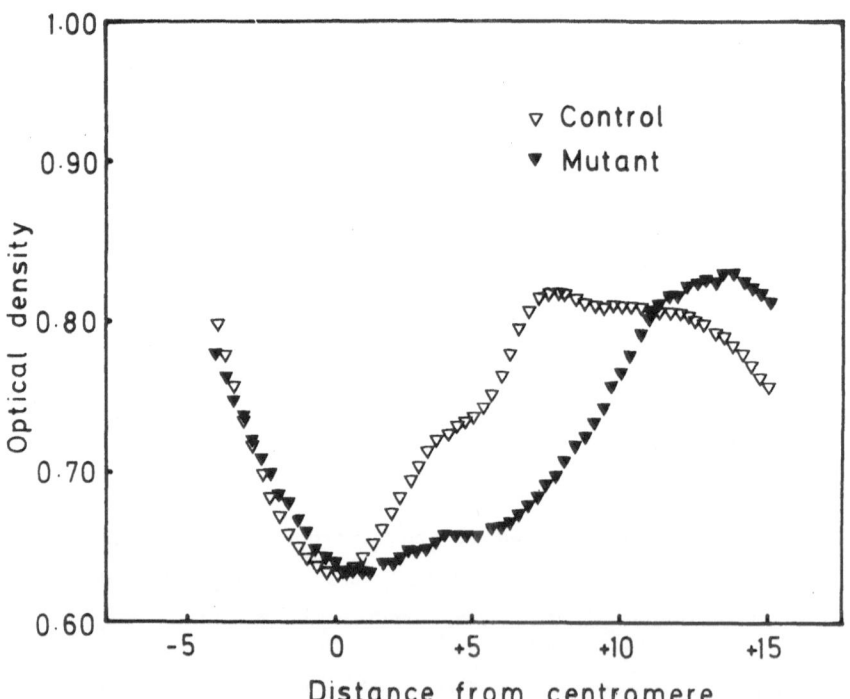

Figure 2. Densitometric profiles of the centrometric region
of chromosome I in the control and in the mutant.
Distances from centromere are % of chromosome length.

DISCUSSION

The M_1 variegated plant which originated the oblong-leaflet mutant studied in this research was characterized by a dicentric chromosome (2n = 10 + 1 telocentric + 1 dicentric), the origin of which and its behaviour were previously described (Errico et al., 1984). The phenotypic analysis carried out on M_2, M_3 and M_4 plants coming from such an M_1 plant evidenced some mutations concerning the leaf and plant structure, the flowering time and the pod fertility (Conicella et al., in press). The oblong-leaflet mutant was identified in M_2. In M_3 this plant segregated 15 oblong-leaflet mutants (two of which were variegated) out of 24 analyzed plants; the mutation bred true in M_4 and M_5 generations.

As no anomalies in the mitotic and meiotic cycles were ascertained in plant $M_3 - 1$ and in its progeny, a karyotypic analysis was performed on these plants in order to ascertain if new chromosome rearrangements were obtained.

It is known, in fact, that the unstable chromosomes can originate new karyotypes through either the disjunction of the associations with their homologous chromosomes or through the classical "breakage-fusion-bridge" cycle. In peas, aneuploids and plants with duplicated chromosome segments were found in the progenies of variegated plants with dicentric chromosomes (Saccardo et al., 1970; Grillo et al., 1981; Monti et al., in press). In Vicia faba, some chromosome mutants were already identified in the progenies of the same M_1 variegated plant which originated the oblong-leaflet mutant (Conicella et al., in press).

The microdensitometric analysis set up in our laboratory on Vicia faba was utilized for the karyotypic analysis of this mutant; this technique allows both a more accurate measurement of the chromosomes and an analysis of their densitometric profiles (Filippone et al., in press).

Data from this analysis showed a longer chromosome 1 in the mutant karyotype. As no difference between the mutant and the control was found in the arm ratio of this chromosome, a careful analysis was performed on its centromeric region: from this analysis it emerged that the difference in the length of chromosome 1 depended on length difference of its centromeric region.

One mechanism hypothesized to explain the origin of the dicentric chromosome in our material is a breakage of one chromosome at the centromere with the formation of two telocentrics and a deletion in another chromosome; the dicentric chromosome originates from the fusion of one of the two telocentrics with the deleted chromosome (Errico et al., 1984). The previous analysis on the variegated M_1 plant which originated the oblong-leaflet mutant showed that the chromosome 1 is certainly involved in the dicentric formation. Our results seem to indicate that the oblong-leaflet mutant is due to the breakage-fusion-bridge mechanism of the induced dicentric chromosome. The longer centromeric region that we have found could be the residue of the intercentromeric region of the dicentric chromosome. A reduction of the intercentromeric region was reported by Hair (1953) in Agropyron which has as a consequence the stability of the dicentric chromosome. Another explanation of the longer length of chromosome 1 of the mutant is that a duplicated segment occurred, as it was demonstrated in pea mutants derived from variegated plants (Grillo et al., 1981; Monti et al., in press).

Other investigations are in course for a finer analysis of the chromosome segment involved in the mutation.

REFERENCES

Conicella, C., Errico, A., Saccardo, F. and La Gioia, N. Variability induced by a dicentric chromosome in Vicia faba L. "E.E.C. Faba Bean Seminar", 13-16 Sept., Nottingham, in press.

Errico, A., Saccardo, F., Monti, L.M. and Conicella, C. 1984. Induction and behaviour of a dicentric chromosome in Vicia faba L. Z. Pflanzenzuchtg, 92, 190-197.

Filippone, E., Smaldone, L.A., Errico, A. and Monti, L.M. 1983. Analisi microdensitometrica di immagini fotografiche di cromosomi di Vicia faba L. Genetica Agraria 37, 169-170.

Filippone, E., Smaldone, L.A. and Monti, L.M. Microdensitometric analysis of Vicia faba L. chromosome images. "E.E.C. Faba Bean Seminar", 13-16 Sept., Nottingham, in press.

Grillo, S. and Saccardo, F. 1981. Induced duplications in chromosome segments of pea and their significance for seed protein studies. E.E.C. Seminar on "Perspectives for peas and lupins as protein group" Sorrento-Sant'Agnello, Oct. 1981.

Hair, J.B. 1953. The origin of new chromosomes in Agropyron. Heredity, Suppl. 6, 215-233.

Monti, L.M. and Saccardo, F. 1969. Mutations induced in pea by X-irradiation of pollen and the significance of induced unstable chromosomes in mutagenic experiments. Caryologia 22, 81-96.

Monti, L.M., Saccardo, F. and Rao, R. Chromosome variation in peas and its use in genetics and breeding. The Pea Crop 40th Nottingham Easter School, in press.

Saccardo, F and Monti, L.M. 1970. Isolation of trisomics in peas. Caryologia 25, 347-358.

Saccardo, F. and Monti, L.M. 1974. Isolation of aneuploids originating from induced unstable chromosome aberrations. IAEVA - Pl - 503/37.

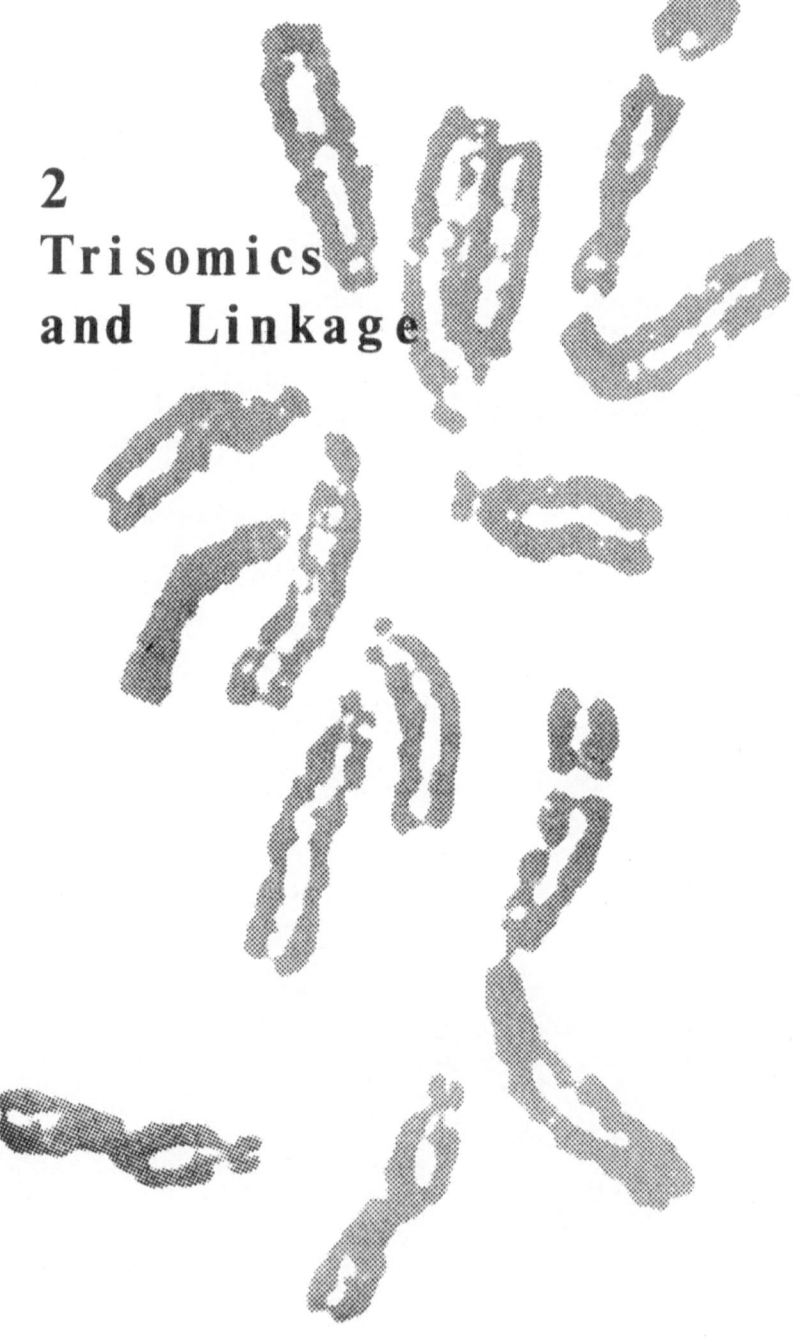

2
Trisomics
and Linkage

THE CYTOLOGY AND MORPHOLOGY OF VICIA FABA TRISOMICS

A. Martin and P. Barcelo

E.T.S.I. Agronomos, Departamento de Genetica, Apartado 3048
Cordoba, Spain.

ABSTRACT

Vicia faba trisomics have been produced by crossing asynaptic mutants with standard genotypes. Only trisomics for the subtelocentric chromosomes were obtained. Giemsa staining technique was used for the identification of the extra chromosome. Two trisomics have been definitely identified. By applying the principal components analysis to quantitative characters from trisomics with similar genetic background, the possible five trisomics for the subtelocentric chromosomes were classified into four phenotypic groups.

INTRODUCTION

Vicia faba is a diploid species with six large chromosome pairs, one satellited metacentric, double the length of the remaining five which are subtelocentric and roughly of the same size. Despite being a well known subject in fundamental cytology, fertile chromosome numeric variants were not described until recently (Poulsen and Martin, 1977; Martin, 1978), while other modes of karyotype variability have been thoroughly studied (Sjodin, 1971 a, b; Schubert et al., 1982).

The first fertile primary trisomics and tetraploids arose on Sjodin's (1971 c) PO-1 mutant (Poulsen and Martin, 1977). A selection of the smaller seeds on the V. faba tetraploid progeny gave rise to some trisomics and double trisomics (Martin, 1978).

The double trisomics could not be maintained. Following Michaelis and Rieger (1959) nomenclature, all trisomics, except one, were for chromosome VI (Gonzalez-Garcia et al., 1981; Schubert et al., 1982). Further effort for producing trisomics by means of the tetraploid failed, since the cross between tetraploid and diploid was never obtained (Martin, 1978).

Sjodin (1970) obtained asynaptic mutants in V. faba and looked for aneuploids in the progeny of asynaptics pollinated with normal plants. However, no aneuploids were found. After selecting high asynaptic plants from Sjodin material and pollinating them with standard material, aneuploids were obtained with high frequency (Gonzalez-Garcia and Martin,

1983). On the other hand, Schubert et al. (1983) crossed two different translocation lines, each involving the same chromosome, and produced definite aneuploids.

Our purpose is to obtain the complete series of V. faba primary trisomics with the final aim of mapping genes. Consequently after a gene has been located on a specific chromosome secondary trisomics will be helpful to establish its position with respect to the centromere or the telomere. We are therefore maintaining the secondary aneuploids which appear in the progeny of primary trisomics.

Using the Giemsa staining technique every individual chromosome has its pattern (Greilhuber, 1975; and many others) and all the possible trisomics can be identified. However, this identification method is a time-consuming task, and for some trisomic lines, seed manipulations or taking roots from pots could damage the plant. A key for phenotypic classification could be a great help for manipulating trisomics.

We have produced over 100 trisomics in the last three years and trisomics for chromosome I have never been obtained. On the contrary trisomic for chromosome VI is the most frequently obtained. As might be expected, independently of the pollinator used, some phenotypic groups are evident. Trisomics for chromosome VI were always included in the same phenotypic groups.

We have tried to identify which characters are the most important for defining the V. faba trisomics and the number of groups in which trisomics can be classified.

MATERIALS AND METHODS

Four Vicia faba lines differing in the asynaptic level were kindly supplied by Dr. Jan Sjodin, Svalof, Sweden.

Three normal lines for meiotic behaviour differing in qualitative and quantitative characters were used as pollinators. In table 1, some of their characterstics are shown. All of these lines were self-fertile and well adapted to green house conditions. Line VFM 15 was used as earliness and white flower donor. Line VFM 27 conferred to the resulting trisomics, disease resistance, self-fertility and reduced seed size. Line VFM 61 gives high yielding hybrids when crossed with the asynaptic mutant.

Line	Origin	Flower colour	Seed size	Seed colour	Generations of selfing
VFM 15	Ethiopia	White	Medium	White	5
VFM 27	Afghanistan	Standard	Small	Black	5
VFM 61	U.K.	Standard	Small	Black	3

Table 1. Characters of Vicia faba lines used as pollinators of asynaptic mutants.

The trisomics were obtained and maintained in a green-house in which temperature varied between 10-35°C and humidity between 50-90%.

In order to obtain trisomics, two flowers per raceme on synaptic plants were emasculated and one or two days later (depending on environmental conditions), they were pollinated with one of the three non-asynaptic plants previously described.

Principal component analysis (Kendall, 1972) was applied to the progeny obtained from the crossing of one single asynaptic plant with VFM 15 line. The asynaptic line had been selfed at least four times. 126 seeds were sown at the same time in individual pots full of sandy soil. For assessing chromosome number, root tips were taken from the roots that grew out of the bottom of the pot. They were treated with a colchicine solution (0.05%), at 20°C in the dark for 3 hours, then fixed in ethanol:acetic acid (3:1) and stained by the Feulgen procedure. Next, buds were taken from each plant at the same plant level and number for cytological studies. They were then fixed and stained following the same procedure.

A Fusarium disease affected all the plants on the experiment. The trisomics were affected more severely than the diploids.

Characters used in principal component analysis:

1. Emergence (days from sowing)
2. Three leaf stage
3. Four leaf stage
4. Days to flowering
5. Days to first mature pods
6. Flowers per node

7. Pods per node

8. Ovules per ovary

9. Nodes to the first flower

10. L: leaflet length

11. Sl: leaflet width at 3/4 from base

12. A2: leaflet width at 1/2 from base

13. A3: leaflet width at 1/4 from base

14. Maximum leaflet width/distance to leaflet end

15. I1 = Al/L

16. I2 = A2/L

17. I3 = A3/L

18. Seeds produced

19. Mean seed weight

20. Mean seed weight of well formed seeds

21. Seed length in cm.

22. Seed width in cm.

23. Seed thickness in cm.

The following modifications were added to the procedure of Greilhuber (1975) for chromosome banding.

Pretreatment: Root tips about 1 cm. long were treated as stated before for assessing chromosome number.

Storage: Roots were used immediately or stored in absolute ethanol at 4°C.

Hydrolysis: This is a critical step. The banding pattern can be modified by changing time. Hydrolysis was in 0.2 M HCl for $2^{1}/2$ hours at room temperature (about 25°C).

Incubation: A saturated solution was added of Barium hydroxide at 45°C for 5.5 minutes. The material was then rinsed by running tap water. The slides were then incubated with 2 x SSC at room temperature for 15 minutes, and afterwards with 2 x SSC at 52°C for 90 minutes.

Pollen fertility was estimated from microscopic analysis of pollen stained with aceto carmine.

RESULTS

Trisomic production

As stated in Material and Methods, the asynaptic lines differed on a pairing level scored as univalent frequency at metaphase I. Our attempt to produce aneuploids succeeded when lines with high asynaptic level (6 univ/cel) were used but failed when asynaptics with a low level were employed. Furthermore, the pollen parents did not affect the outcome. We have observed, as did Sjodin (1970), a variation for the same line of asynaptic levels in different years. However, when asynaptics were crossed with normal plants and the F_2 were analysed the asynaptics obtained were roughly of the same level. Consequently, it could be inferred that differences of asynaptic level are not the result of modifiers but of environment, since it is unlikely for a normal diploid to carry exactly the same modifiers as the selected one for an asynaptic mutant. In addition, we have found decreasing univalent frequency with inbreeding level (unpublished results).

In 1981 we succeeded in producing primary trisomics and double trisomics. We crossed asynaptics differing in pairing levels with standard lines. In 96 seeds from the line with intermediate asynaptic level 9 trisomics were obtained. In 75 seeds from a stronger asynaptic line 9 trisomics and 1 double trisomic were produced. The higher aneuploids frequency has been obtained when the asynaptic parental shows the higher univalent frequency. For instance, on the material in which principal components analysis was applied 23 trisomics from 126 seed were obtained, that is 18.2% of aneuploids. The maternal plant had 6.6 univalents per cell.

Also, aneuploids appear in selfed asynaptics. However, the viability and fertility of these plants are much lower than those of the aneuploids obtained by outcrossing asynaptics.

Morphology

Four phenotypic groups have been observed based on leaflet morphology (Fig. 1-5).

In group 1 two different phenotypes are apparent when lines 27 and 61 are used as pollinator (Figs. 1 and 2). We believe that two different trisomics are included in this group.

The phenotypic aspect is kept from one generation to the next. However, on two occasions this rule has not been fulfilled. One trisomic for phenotypic group 2 (Fig. 3) and another for group 3 (Fig. 4) appeared on the progeny of trisomic for group 1. In both cases the trisomics were not asynaptics. They could not come from an asynaptic trisomic and for this reason they probably are the result of trivalent drift.

1   2

Figures 1 and 2. Two different <u>Vicia</u> <u>faba</u> trisomics
included in phenotypic group 1. Leaf aspect.

3   4

Figure 3.   Leaf from phenotypic group 2 of <u>Vicia</u> <u>faba</u> trisomics.

Figure 4.   Leaf from phenotypic group 3 of <u>Vicia</u> <u>faba</u> trisomics.

5   6

Figure 5. Leaf from phenotypic group 4 of <u>Vicia</u> <u>faba</u> trisomics.

Figure 6. Leaf from <u>Vicia</u> <u>faba</u> diploid.

After the former difficulty appeared, and given the practical impossibility of using Giemsa banding technique in every F_1 for mapping purposes, it was clear to us that a phenotypic classification was necessary as well as the determination of the influence of different morphological characters in this classification.

We applied the principal component analysis (P.C.A.) to the largest progeny from one asynaptic plant pollinated with line VFM 15. As both parental lines were quite homozygous we can assume that the observed morphological differences were due to the extra chromosome, since every plant had a similar genetical background.

In figure 7 we present the grouping of 32 individuals on the two axes obtained after principal component analysis was applied. These two axes explain 60% of variation and classify the 23 trisomics and 9 euploids into five phenotypic groups.

There were five plants which could not be included in any group. They were atypical due fundamentally to a bad development or because they suffered more severely from the infection with <u>Fusarium</u> which affected reproductive characters. Plants 30 and 32 were diploids. Plant number 3 is identical on leaflet morphology to plant 13. Plants 16 and 17 were classified on banding pattern as trisomics for chromosome V, their progeny were included in phenotypic group 3.

The more determinant characters on defining the first two axes were: 11, 12, 13 and 5 for the first one, and 15, 17, 16 and 20 for the second, namely leaflet indices. Vegetative period (5) and seed weight (20) were also important.

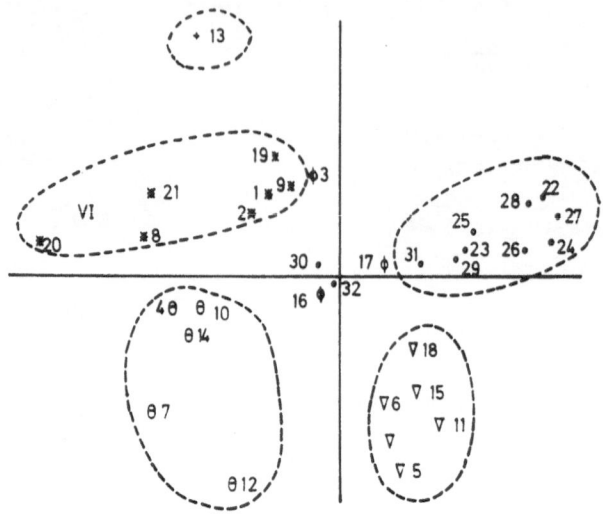

Figure 7.   Grouping of 32 individuals on two axes after
   principal component analysis.

On the basis of our previous observation of leaf morphology and after
the P.C.A. (fig. 7) we have classified V. faba trisomics in four
phenotypic groups.   Group 1 includes plants 20, 21, 8, 2, 1, 9 and 19.
All of them were trisomic for chromosome VI.

   Group 2 consisted of plants 13 and 3. They are the most sensitive to
environmental conditions.

   Group 3 belong to the group of plants 5, 6, 11, 15, 16, 17 and 18.
All of them were trisomic for chromosome V.

   Group 4 includes plants 4, 7, 10, 12 and 14.

Cytology

   In figure 8 we present our Giemsa banded karyotype Vicia faba, and in
figure 9 using the same technique, a trisomic for chromosome VI is shown.

   We have found some mitotic irregularities in some V. faba trisomics
but this aspect will not be discussed here. In table 2, results from the
analysis of metaphase I in plants for the four phenotypic groups are
presented. In very low frequency four bivalents, two univalents and one

trivalent are found. This configuration, if present on the mother egg cell could give rise to trivalent drift.

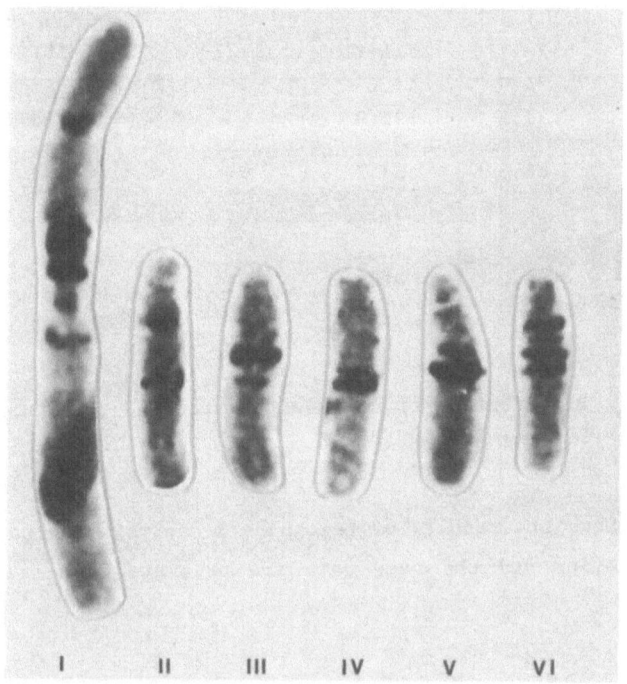

Figure 8. _Vicia faba_ Giemsa-banded karyotype.

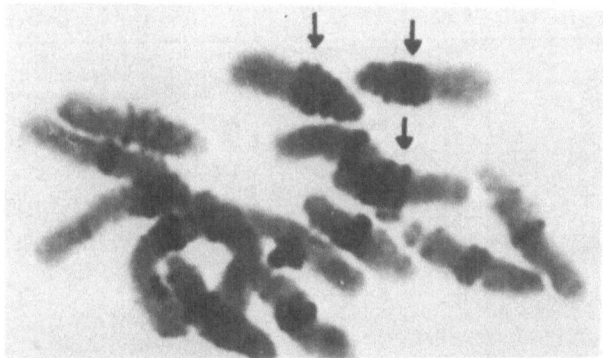

Figure 9. _Vicia faba_ Giemsa-banded trisomic for chromosome VI (arrowed).

Phenotypic group	Type of configuration				Number of cells
	511+1111	611+11	511+31	411+21+iiii	
1	74.2	25.9	0	0	204
2	65.3	34.7	0	0	225
3	60.8	37.3	1.5	0.5	675
4	68.2	31.8	0	0	415

Table 2.   Distribution of cells with different configurations in metaphase I. (percentages)

Table 3 shows the results of telophase I analysis for each phenotypic group of trisomics and the same data are presented for telophase II in table 4.

Phenotypic group	Micronuclei per dyad				Number of dyad
	0	1	2	3	
1	88	11.3	0.7	0	1254
2	85.6	13.5	0.8	0	400
3	82.6	13.2	4.1	0.1	936
4	91.7	8.0	0.1	0.1	675
Diploid	99	1	0	0	400

Table 3.   Frequencies of different micronuclei number at anaphase I (in percentages)

Phenotypic group	Number of micronuclei					Number of tetrads
	0	1	2	3	>3	
1	86.3	11.1	2.5	0.1	0.0	2100
2	77.7	13.0	8.0	1.0	0.3	300
3	78.1	13.0	8.0	0.4	0.4	1500
4	94.2	4.6	1.2	0.0	0.0	1500
Diploid	97.0	2.0	1.0	0.0	0.0	300

Table 4. Frequencies of different numbers of micronuclei in per cent at tetrad stage.

In table 5 we present data on pollen fertility and transmission of trisomy from the progeny of plants in which principal component analysis was applied. Additionally, mean values and standard deviations are presented in table 6. Atypical trisomics 3, 16 and 17, and diploids 30 and 32 have been disregarded.

Phenotypic groups	Pollen grain counted	Normal pollen	Seeds controlled	Trisomics
1	4200	43.1	98	20.4
2	600	35.0	52	7.7
3	600	65.2	134	15.7
4	3000	48.5	29	31.0
Diploid	600	93.7	-	-

Table 5. Pollen fertility and transmission rate of Vicia faba trisomics in %.

Characters

Phenotypic group	First Pod	A_1	A_2	A_3	A_1/L	A_2/L	A_3/L	Seed weight(g)	No. of Plants
1	55.±3.2	1.77±0.1	2.69±0.2	2.54±0.2	0.34±0.01	0.51±0.01	0.48±0.01	0.74±0.03	6
2	47	1.76	2.43	2.15	0.26	0.36	0.32	1.07	1
3	47.4±0.2	2.74±0.6	3.46±0.1	3.00±0.0	0.61±0.01	0.78±0.02	0.67±0.02	0.57±0.03	5
4	58.6±1.0	1.76±0.1	2.17±0.2	1.88±0.2	0.68±0.06	0.74±0.04	0.63±0.03	0.68±0.05	5
Diploid	39.7±0.6	3.16±0.1	4.37±0.2	3.73±0.1	0.49±0.01	0.68±0.01	0.58±0.01	0.76±0.01	9

Table 6. Mean values from characters defining the two first axes after principal component analysis, (explanation in the text).

DISCUSSION

The principal component analysis applied to a genetically homogeneous group of diploids and trisomics V. faba shows that leaflet morphology is the best character for defining trisomics. This makes it possible to distinguish trisomics from each other and from the disomic sibs. In table 6 days to first pod and seed weight also appear as playing a role in this differentiation. However, the first pod is only useful to identify diploids against all trisomics together, and seed weight distinguishes trisomics of phenotypic group 2 (fig.3) which have a distinct leaflet and stem aspect.

We believe that in the analysed group, one trisomic has been missed since in other crosses two distinct trisomics are included on phenotypic group 1. This assumption remains to be worked out, but if true, V. faba should be included in the species group in which trisomics are morphologically distinct to diploids and can be distinguished from each other phenotypically (Khush, 1973). In addition, the number of trisomics obtained in a single female plant is not large, therefore a particular trisomic can be missed. Again, trisomics are not produced at the same rate. Trisomics for chromosome VI have been obtained more frequently than any other and this chromosome is the shortest (Sjodin, 1971 b). In the group studied, trisomics for chromosomes V and VI appeared 7 times each, nearly 2/3 of the total number of trisomics obtained and chromosome V is the shortest after chromosome VI. That is, a negative correlation seems to exist between chromosome length and trisomic frequency. Consequently, it should be expected that trisomic for chromosome II is the one missed in the group to which principal component analysis was applied, since it is the longest of sub-telocentric chromosomes.

In table 2, no clear relationship between chromosome length and trivalent frequency can be observed, since trisomics for chromosomes VI and V (the smallest ones) show the highest and lowest trivalent frequency respectively. The chromosome associations with 3 univalents could account for the production of unrelated trisomics, supposing that 2 univalents are from another chromosome different than the critical in trisomy. Further, we have to assume that this situation is present on mother egg cell since transmission by male, if it exists, is rare.

Neither a clear relationship exists between trivalent frequency and micronuclei at telophase I or II (tables 2, 3 and 4) but the two groups with the lowest trivalent frequency present more micronuclei and the transmission rate of trisomy by female is lower in both groups (table 5). It can be inferred therefore that other factors in addition to chromosome length influence meiotic behaviour and transmission rate.

After observing the low transmission of trisomy in some groups and the decrease of fertility and vigour associated to them, we are now using parentals VFM 27 and VFM 61, with smaller seeds. With these parentals we obtain larger and small seed sized progenies. The trisomics obtained in this way produce more seed, so we have increased the chance of maintaining them. On the other hand these small seed sized trisomics are essential to study characters associated with large seed sized plants.

ACKNOWLEDGEMENTS

The senior author is grateful to Dr. Sjodin for providing seed samples of asynaptic Vicia faba. We would both like to express our gratitude to Professor J.I. Cubero who has shown an unceasing interest in our work and to the Ramon Areces Foundation for financial support. We are also indebted to Dr. G. Chapman for critical reading of the manuscript.

REFERENCES

Gonzalez-Garcia, J.A., Padilla, J.A. and Martin, A. 1981. Characteristics of Vicia faba trisomics. FABIS 3, 30.

Gonzalez-Garcia, J.A. and Martin, A. 1983. Development, use and handling of trisomics in Vicia faba. FABIS 6, 10-11.

Greilhuber, J. 1975. Heterogeneity of heterochromatin in plants: Comparison of Hy- and C- Bands in Vicia faba L. Plant Sys. Evol. 124, 139-156.

Kendall, M.G. 1972. A course in multivariate analysis. Ed. Griffin, London, 185 pp.

Khush, G.S. 1973. Cytogenetics of aneuploids. Ed. Academic Press, New York and London, 301 pp.

Martin, A. 1978. Aneuploidy in Vicia faba. The Journal of Heredity. 69, 421-423.

Michaelis, A and Rieger, R. 1959. Strukturheterozygotie bei Vicia faba. Zuchter, 29, 354-361.

Poulsen, M.H. and Martin, A. 1977. A reproductive tetraploid Vicia faba L. Hereditas 87, 123-126.

Schubert, I., Michaelis, A. and Rieger, R. 1982. Karyotype variability and evolution in Vicia faba L. Biol. Zbl. 101, 793-806.

Schubert, I., Rieger, R. and Michaelis, A. 1983. A method for direct production of definite aneuploids of Vicia faba L. FABIS 7. 13-18

Sjodin, J. 1970. Induced asynaptic mutants in Vicia faba L. Hereditas 66, 215-232.

Sjodin, J. 1971 a. Induced paracentric and pericentric inversions in Vicia faba L. Hereditas 67, 39-54.

Sjodin, J. 1971 b. Induced translocations in Vicia faba L. Hereditas 68, 1-34.

Sjodin, J. 1971 c. Induced morphological variation in Vicia faba L. Hereditas 67, 155-180.

3
Disease
Resistance

A NOTE ON THE I.L.B. SOURCE OF BOTRYTIS FABAE RESISTANCE

L.D. Robertson

ICARDA, P.O. Box 5466, Aleppo, Syria.

ICARDA has screened a large proportion of its germplasm collection for chocolate spot resistance and two selections, ILB 438 and ILB 938, were located. ILB 438 is an accession that came to ICARDA from Ecuador through Columbia and ILB 938 is a mass selection from ILB 438 based on seed size. It is of considerable interest that the source is an area outside of the presumed original area of domestication in the Old World and has a relatively small area under faba bean cultivation. Selections have been found with multilocational resistance to chocolate spot in Canada, Egypt, Syria and the United Kingdom.

These selections from ILB 438 and ILB 938 have been distributed to breeders and pathologists in Canada, Europe and the Middle East. Crosses have been made at ICARDA and in Egypt and promising lines for both yield and disease resistance obtained. At ICARDA lines have been found with yields as good as the best control and with high levels of chocolate spot resistance. ICARDA is currently purifying and increasing several derivatives from ILB 438 and ILB 938 with uniformly high resistance. New collections have been obtained from the Columbia-Ecuador border region and will shortly be evaluated for disease resistance.

Editor's Note

This material is of great interest since it originates outside the presumed area of domestication and apparently confers some resistance against more than one disease as shown in the following paper. It presents, too, a problem in strategy if it is to be uniformly adopted by breeders.

IDENTIFICATION OF SOME SOURCES OF RESISTANCE FOR CHOCOLATE SPOT AND RUST IN FABA BEANS

S.A. Khalil[1], A.M. Nassib[2], H.A. Mohammed[3] and W.F. Habib[4]

1 and 2 Food Legumes Research Section
(Field Crops Research Institute)

3 and 4 Legumes Research Pathology Section
(Pathology Research Institute)

A.R.C., Giza, Egypt.

ABSTRACT

The present studies were carried out under ICARDA/IFAD Nile Valley project in Egypt, during the three winter seasons, 1980/81, 1981/82 and 1982/83 to detect the field resistance of some faba bean breeding lines to chocolate spot (B. fabae) and rust (U. fabae), where both are the main destructive diseases of faba bean in North Delta region. The field experiments were conducted under natural epiphytotic conditions at Sakha and Nubaria and controlled conditions at Giza Research Stations.

INTRODUCTION

Faba bean (Vicia faba L.) is the oldest pulse crop grown for seed in Egypt, due to its high nutritive value and its role as a break crop in cereal rotation system.

The average cultivated area ranged from 104794 ha., in 1979 to 121603 ha. in 1983 season, with an average of 108832 over the last five years. This area is located under three ecological regions: North Delta, Middle and Upper Egypt, and distributed to 29.4%, 46.0% and 24.6% for the three regions, respectively.

The average seed yield per unit area tended to be the lowest (1.9 T/ha) in North Delta compared to those of Middle Egypt (2.2 T/ha) and Upper Egypt (2.6 T/ha), mainly due to foliar disease incidence, i.e. chocolate spot (B. fabae) and rust (U. fabae) diseases.

The present studies were conducted under ICARDA/IFAD Nile Valley project and partially supported by the International Development Research Centre (IDRC) in Egypt, to detect the field resistance of some faba bean breeding lines to chocolate spot (B. fabae) and rust (U. fabae) where both are the main destructive diseases of faba bean in North Delta Region.

MATERIALS AND METHODS

The present studies were carried out at Sakha, Nubaria (North Delta) and Giza research stations, during the three growing seasons 1980/81, 1981/82 and 1982/83 to identify faba bean breeding lines for : (A) Chocolate spot (B. fabae) and (B) Rust (U. fabae) resistance.

The field studies were carried out under the natural epiphytotic conditions at Sakha and Nubaria (disease prevailing area) and under controlled conditions at Giza (disease free area) research stations. The evaluation was also extended to cover the potted plant experiments under the glasshouse conditions at Giza Research Station.

(A) Field experiments

1980/81 Season. Thirty four breeding lines compared with the recommended variety Giza 3 and the highly susceptible variety Rebaga 40 were tested in a randomized complete block design with two replicates at Sakha research station.

The experiment was replicated twice where one experiment was protected against disease infection by spraying with Dithane M 45, and the other experiment was left without spraying.

Dithane M 45 was sprayed four times with the recommended dose when foliar diseases began to appear (1% infection), and untreated plots were sprayed with water.

1981/82 season. Ten promising genotypes along with the two mentioned commercial cultivars (G.3 and R.40) were grown in a split plot design, with three replicates where the main plots were assigned to Agrimycin (sprayed or non-sprayed) as plant protection treatments, and the sub-plots to the tested genotypes. The experiment was conducted at Sakha and Nubaria Research Stations.

1983/83 season. Six genotypes compared with the two commercial varieties were tested in a split plot design with three replicates, where the main plots were assigned to (a) artificial inoculation with a single chocolate spot isolate (150×10^3 conidia/ml), (b) uninoculation, and the sub-plots to the tested genotypes. The whole experiment was conducted at Sakha and Nubaria, and controlled conditions at Giza Research Stations. To ensure the successful evaluation, the plant material 80 days old (at Giza) were adequately exposed to the most aggressive chocolate spot isolate (Nubaria), and then covered with polyethylene sheet to maintain leaf wetness over night, as described by Wilcoxson et al. (1975), Parlevleit and Van Ommeren (1975), Conner and Bernier (1982b), Bernier

(1983) and Hannonik (personal communication).

The experimental plots were 3.6, 5.4 and 5.4 m^2 during 1980/81, 81/82 and 82/83, respectively. All the recommended cultural practices were followed in the field test.

B. Potted plant experiments

The studies were carried out under the glasshouse conditions at Giza Research Station, to evaluate the same genotypes with both potted plant experiments and detached leaves method.

In 1981/82 season. The ten genotypes, included in the field test along with Giza 3 and Rebaya 40 were grown in pots where two experiments were conducted, each with three replicates.

When plants were 40 days old detached leaves and the whole plants of the first experiment were artificially inoculated with a spore-suspension (175×10^3) conidia/ml) of chocolate spot disease. The second experiment was inoculated when plants were 80 days old.

In 1982/83 season. The artificial inoculation with a spore suspension (150000 conidia/ml) of a single aggressive isolate from Nubaria was applied on detached leaves by using the micro meter syringe (Khalil and Harrison 1981) when pottled plants were 80 days old.

Varietal reaction to chocolate spot disease was recorded in the field test as percentage; however, in the potted plant experiments, the scale 0-9 was used, and on detached leaved method, the average dimension of the inoculated sites was measured in mm.

Reaction to rust disease was recorded as percentage, in the field test at Sakha and Nubaria Research Stations.

RESULTS AND DISCUSSIONS

(A) Chocolate spot (B.fabae) disease

Field experiments:

In 1980/81 season. Records on chocolate spot disease at Sakha Research Station (Table 1) indicated that the infection was light until the end of February varying from 0.0 to 1% starting 6th March, infection increased gradually to reach its maximum on 15th April.

Av. % of leaf spot infection on

Lines/cultivars	Feb. 21		March 6		March 20			April 25		
	NS	S	NS	SP	NS	SP	mean	NS	SP	mean
Giza 1	0	0	5.0	5.0	15.0	7.5	11.2	30.0	10.0	20.0
Giza 3	trace	trace	3.0	3.0	25.0	10.0	17.5	37.5	15.0	26.2
Rebaya 40	1	1	10.0	5.0	40.0	27.5	33.7	40.0	40.0	40.0
122/27/78	0	0	5.0	1.0	10.0	5.5	7.7	22.5	15.0	18.7
147/125/78	0	0	10.0	10.0	27.5	10.0	18.7	30.0	7.5	18.7
148/3515/78	0	0	5.0	3.0	22.5	5.0	13.7	25.0	7.5	16.2
138A/2143/77	0	0	7.1	1.0	32.5	5.0	18.7	40.0	7.5	23.7
187/2324/79	0	0	5.0	1.0	40.0	5.0	22.5	30.0	7.5	18.7
249/801/80	0	0	3.0	1.0	5.0	1.0	3.0	13.5	5.0	8.7
249/802/80	0	0	1.0	1.0	7.5	1.0	4.2	5.0	3.0	4.0
249/803/80	0	0	1.0	1.0	3.0	1.0	2.0	10.0	5.0	7.5
249/804/80	0	0	1.0	1.0	10.0	1.0	5.5	10.0	3.0	6.5
258/810/80	1	0	5.0	1.0	15.0	5.0	10.0	20.0	10.0	15.0
112/3200/74	trace	0	5.0	3.0	40.0	10.0	25.0	40.0	12.5	26.2
122/65/67	trace	0	5.0	5.0	20.0	5.0	12.5	35.0	7.5	21.2
130/1881/76	0	0	5.0	1.0	17.5	5.0	11.2	20.0	10.0	15.0
133/2067/77	0	0	7.5	3.0	30.0	5.0	17.5	35.0	12.5	23.7
80-B/2528/70	0	0	3.0	3.0	17.5	10.0	13.7	20.0	15.0	17.5
90/1966/72	0	0	15.0	7.5	25.0	10.0	17.5	35.0	10.0	22.5
Protein 10/78	0	0	5.0	1.0	25.0	5.0	15.0	40.0	10.0	25.0
" 56/78	1	1	15.0	10.0	45.0	20.0	32.5	40.0	20.0	30.0
" 88/78	1	1	7.5	1.0	17.5	5.0	11.2	22.5	7.5	15.0
" 114/78	0	0	5.5	1.0	27.5	15.0	21.2	40.0	10.0	25.0
Comp. 2 S₂L.29	0	0	7.5	5.0	25.0	7.5	16.2	30.0	25.0	27.5
" " L.32	0	0	5.0	1.0	10.0	5.0	7.5	20.0	7.5	13.7
" " L.58	0	0	7.5	1.0	22.5	3.0	12.7	40.0	5.0	22.5
H.B.P. 3 S₁ L.18	0	0	10.0	1.0	25.0	7.5	16.2	30.0	7.5	18.7
H.B.P. 6 S₁ L.1	0	0	15.0	1.0	22.5	7.5	15.0	30.0	10.0	20.0
H.B.P. 12 S₁ L.80	0	0	5.0	1.0	15.0	7.5	11.2	25.0	10.0	17.5
B.W.C. /503/80	0	0	10.0	5.0	32.5	15.0	23.7	32.5	15.0	23.7
B.W.C. /523/80A	0	0	5.0	1.0	15.0	1.0	8.0	20.0	5.0	12.5
R.C. 39/80	0	0	5.0	1.0	7.5	3.0	5.2	15.0	3.0	9.0
78 S 49456	0	0	1.0	1.0	20.0	1.0	10.5	20.0	3.0	11.5
Seville Giant	0	0	1.0	1.0	10.0	5.0	7.5	10.0	7.5	8.7
NEB 2727/75	0	0	5.0	1.0	30.0	5.0	17.5	35.0	5.0	20.0
NEB 938	0	0	1.0	1.0	10.0	1.0	5.5	10.0	5.0	7.5
Mean	0.1	0.1	5.9	2.5	21.2	6.8	13.9	26.6	10.0	18.5
L.S.D. 5%	-	-	-	-	-	-	9.16		11.29	9.19

NS : not sprayed

S : sprayed with Diathane M 45

Table 1.   Average % of leaf spots on faba bean entries sprayed
or not with Dithane M 45, Sakha Res. Station, 1980/81.

The tested genotypes significantly differed in their reaction to
chocolate spot disease. Lines 249/801/80, 249/802/80, 249/803/80,
249/804/80, Seville Giant, R.C. 39/80 and ILB 938 showed the lowest
percentage of infection. On the other hand lines 139 A/2143/77,
112/3200/74, 122/65/67, 133/2067/77, 90/1966/72, Protein 110/78, protein
56/78, protein 114, Comp 2 S₂, L.29, L.58 and NEB 2727/75 showed the
highest levels of infection. The reaction of the other lines was
intermediate.

Dithane M45 application significantly decreased chocolate spot
infection but the decrease seemed to differ for the different lines.

Susceptible genotypes, responded better to fungicide application and the

Lines/cultivars	Disease reaction		Initial infection	Disease developmt. Sakha,1981	
	Sakha 79/80	Giza* 80/81		Sprayed	Not sprayed
Giza 1	MR	MR	5	5.0	25.0
Giza 3	MR	MR	3	12.0	34.5
Rabaya 40	S	S	10	35.0	30.0
122/27/28	R	untested	5	14.0	17.0
147/125/78	R	"	10	7.5	20.0
148/3515/78	R	"	5	4.5	20.0
139/ A/2149/77	R	"	7.5	6.5	32.5
187/2324/79	R	"	5	6.5	25.0
249/801/80	R	R	3	4.0	9.5
249/802/80	R	R	1	2.0	4.0
249/803/80	R	R	1	4.0	9.0
249/804/80	R	R	1	2.0	9.0
258/810/80	R	untested	5	9.0	15.0
112/3200/74	MR	"	5	9.5	35.0
122/65/67	S	"	5	2.5	30.0
130/1881/76	MR	"	5	9.0	15.0
133/2067/77	S	"	7.5	9.5	27.5
80-B/2528/70	MR	"	3	12.0	17.0
90/1966/72	S	"	15	2.5	20.0
Protein 10/78	R	"	5	9.0	35.0
" 56/78	R	"	15	10.0	25.0
" 88/78	R	"	7.5	6.5	15.0
" 114/78	R	"	5.5	9.0	34.5
Comp. $2S_2L$ 29	R	"	7.5	20.0	22.5
" " L32	R	"	5.0	6.5	15.0
" " L58	R	"	7.5	4.0	32.5
H.B.P_3S_1 L.18	R	"	10	6.5	20.0
H.B.P_6S_3 L. 1	R	"	15	9.0	15.0
H.B.$P_{12}S_1$L.80	R	"	5	9.0	20.0
B.W.C./503/80	R	R	10	10.0	22.5
B.W.C./523/80A	R	R	5	4.0	15.0
R.C./39/80	R	R	5	2.0	10.0
78 S 49456	R	R	1	2.0	19.0
Seville Giant	R	R	1	6.5	9.0
NEB/2727/75	S	untested	5	4.0	30.0
ILB 938	R	R	1	4.0	9.0

*Under artificial epiphytotic

Table 2.   Average reaction, development of leaf spot diseases and response of faba bean promising lines to Diathane M 45, at Sakha, Nubaria and Giza, 1979/80 and 1980/81 seasons.

Disease development estimated as the difference between the initial and maximum infection recorded on 6th March and 15th April, respectively, was calculated for different genotypes under the two fungicidal treatments (Table 2).   Disease development was less in the sprayed plots compared with the non-sprayed plots.   However, differences in disease development under sprayed and non-sprayed conditions were variable with different lines.   In case of resistant lines, disease development did not show considerable variation under both treatments.   On the other hand, on susceptible lines, Dithane M 45 application was quite effective in reducing disease development.

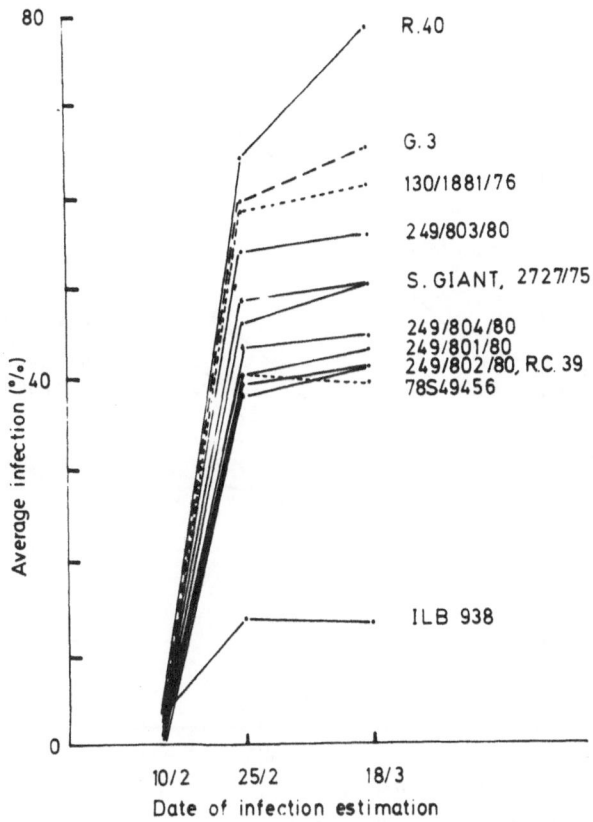

Figure 1. Development of field infection with chocolate spot on twelve faba bean genotypes at Nubaria.

In 1981/82 season. Results of chocolate spot at Nubaria (Fig. 1) showed a low rate of infection in early February, and the infection significantly increased from 10th February to 25th February to 16th March. In all cases spraying with Agrimycin decreased the percentage of infection, this decrease was significant on 25th February and 16th March. At Sakha Research Station infection was mild, the same trend was observed although the differences were not significant.

On the average of both locations ILB 938 proved to be the most promising line.

In 1982/83 season. At Nubaria results showed low rate of chocolate spot infection (Fig. 2) in early February and significantly increased on 8th February through 8th March to 23rd March. The average infection over all entries was significantly 48% higher under artificial inoculation than under natural epiphytotic.

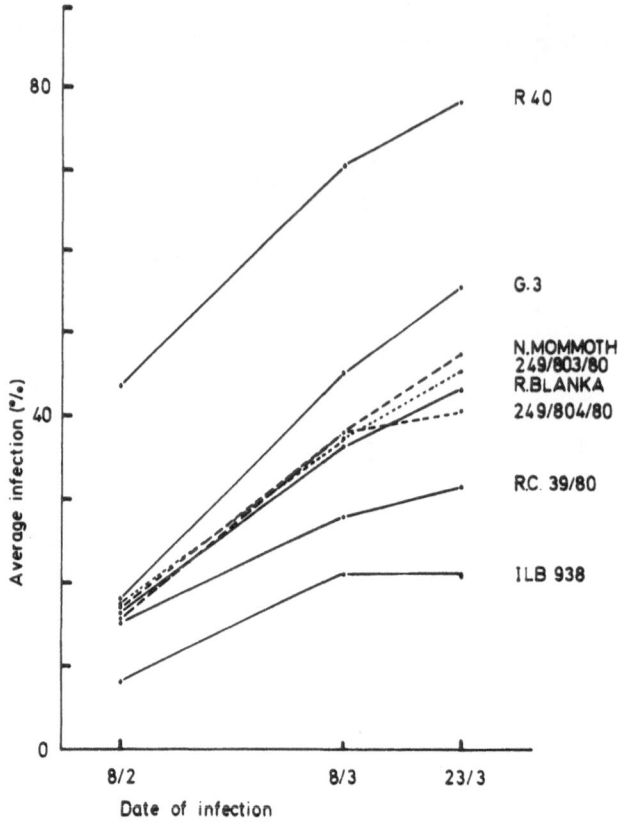

Figure 2.   Development of artificial infection with chocolate
spot on eight faba bean genotypes at Nubaria.

At Sakha Research Station infection was mild (Fig. 3) when recorded
at 12th February, 13th March and 20th March and significantly increased
from 27th March to 6th April.

At Giza Research Station, the inoculated plots recorded the highest
range of infection compared with both other locations and significantly
increased from 7th February through 12, 18 and 15 March (Fig. 4).

Over the three locations, ILB 938 was the most promising line
followed by 249/804/80 and R.C. 39/80 and all were significantly lower
than the check cultivar Giza 3. Over all locations artificial inoculation
with chocolate spot disease increased the infection by about 48%.

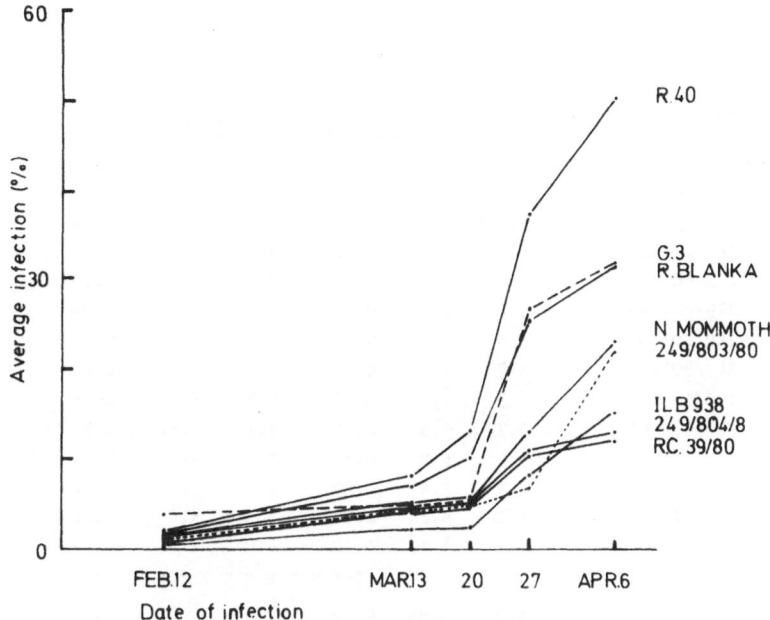

Figure 3. Development of artificial infection with chocolate spot on eight faba bean genotypes at Sakha.

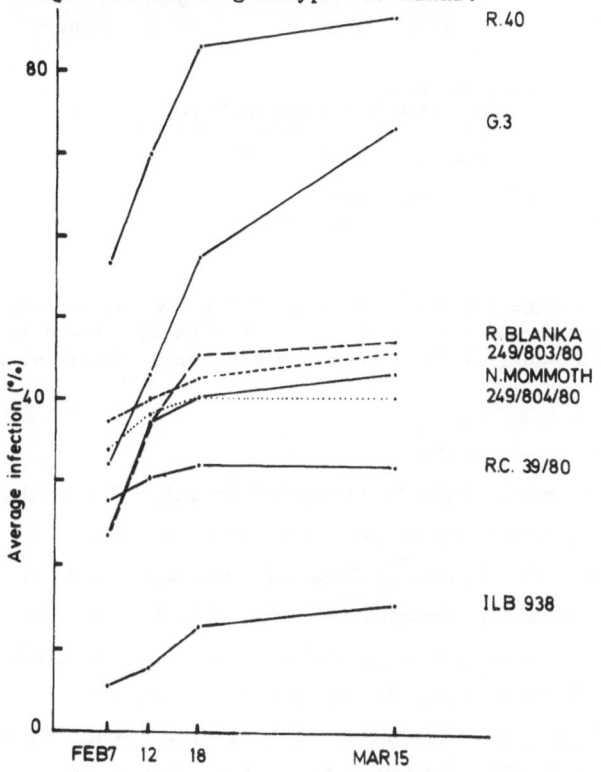

Figure 4. Development of artificial infection in the field with chocolate spot on eight faba bean genotypes at Giza.

| Genotypes | 40 days | | | 80 days | | |
| | Average reaction | | D.R.* | Average reaction | | D.R. |
	48 hours	One week		48 hours	One week	
Giza 3	3.5**	5.1	0.32	4.5	5.9	0.28
Rebaya 40	3.6	5.7	0.42	4.1	5.8	0.34
249/801/80	3.8	5.5	0.34	4.2	5.4	0.24
249/802/80	3.7	5.1	0.28	3.7	3.9	0.04
249/803/80	3.3	4.8	0.30	4.1	5.5	0.28
249/804/80	3.8	5.5	0.34	3.3	3.7	0.08
130/1881/76	3.3	4.8	0.30	4.1	5.5	0.28
Seville Giant	3.7	5.3	0.32	3.7	4.7	0.20
R. C. 39/80	3.4	5.2	0.36	3.8	5.1	0.26
ILB 938	2.4	2.4	0.00	2.5	3.1	0.12
2727/75	3.8	5.2	0.28	4.3	5.6	0.26
78 S 49456	3.7	4.7	0.20	4.0	4.7	0.14
M e a n	3.5	4.94	0.29	3.9	4.9	0.20
L.S.D. 0.05	N.S	0.78		0.63	0.83	
C.V.	12.58	7.62		7.89	2.29%	

* D.R. (Development rate)

$$= \frac{\text{Difference in average reaction}}{\text{Infection period} (= 5 \text{ days})}$$

** Scale : 0 = Highly resistant.
9 = Highly susceptible.

Table 3. Average reaction of faba bean genotypes to chocolate spot disease under controlled conditions at 40 and 80 days after planting, at Giza Research Station, 1981/82 season.

(B) Potted plant experiments

In 1981/82 season. Results presented in Table 3 revealed significant differences among tested genotypes, over the two stages of growth, line ILB 938 recorded the lowest values of infection and rate of disease developments. However, results of the infected plants 80 days old indicated that lines 249/802/80, 249/804/80, Seville Giant and 7854956 were significantly lower than the check cultivar Giza 3.

In 1982/83 season. Results of detached leaf method (Fig. 5) indicated that ILB 938, 249/804/80 and R.C. 39/80 were able to retard chocolate spot disease.

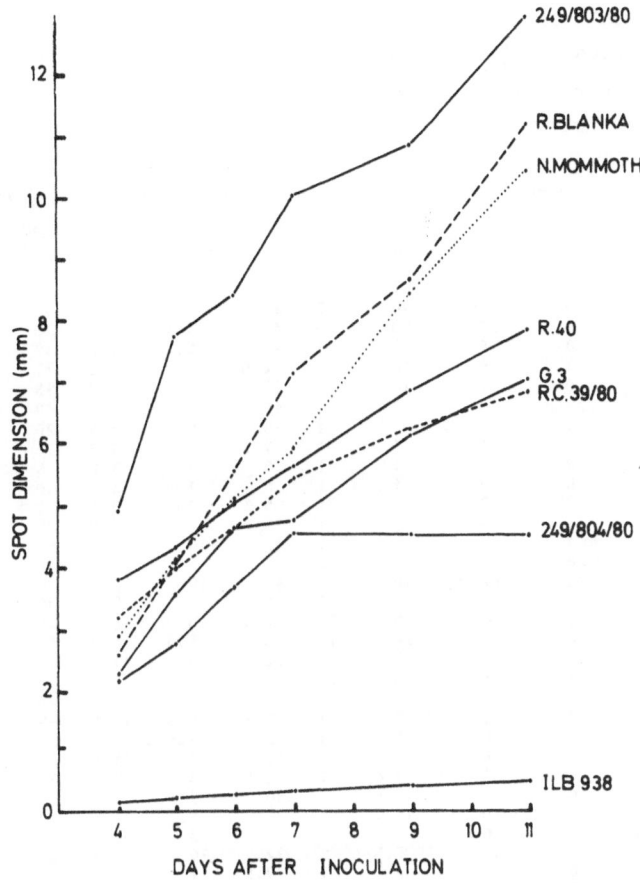

Figure 5.   Development of chocolate spot on inoculated detached
leaves of eight faba bean genotypes, sown in pots at Giza.

From reviewing the results of studied genotypes under the field and controlled conditions, it is clear that line ILB 938 proved to be highly resistant either in the field or controlled conditions. This finding is in accordance with El-Sherbeeny and Mohammed (1980) in Egypt.   On the other hand, this line was rated resistant, at Latakia, Syria (Honnonik, personal communication),   and ranked the top of thirteen tested genotypes at Cambridge, England, U.K. (Jellis et al., 1982).   The results of the field test supported by those of potted plants and detached leaf method give a clear indication that the three lines : ILB 938 followed by lines 249/804/80 and R.C. 39/80 could be considered a source of chocolate spot resistance.

(B) Rust disease (U. fabae)

Rust caused by U. fabae is the second destructive foliar disease of faba beans in North Delta region.

Lines/cultivars	Feb. 21		March 6		March 20			April 15		
	NS	S	NS	S	NS	S	Mean	NS	S	Mean
Giza 1	0	0	7.5	5	22.5	7.5	15	30	10	20
Giza 3	0	trace	10	5	25	10	17.5	35	15	25
Rebaya 40	1	1	20	25	40	27.5	33.7	30	25	27.5
122/27/78	0	0	10	1	17.5	5.5	11.5	30	17.5	23.7
147/125/78	1	0	10	10	22.5	10	16.2	25	15	20
148/3515/78	0	0	10	5	22.5	5	13.7	35	7.5	21.2
139/ A/2143/77	trace	0	10	1	40	3	21.5	40	5	22.5
187/2324/79	1	0	5	1	20	5	12.5	20	10	15
249/801/80	0	0	5	1	7.5	1	4.2	10	5	7.5
249/802/80	0	0	1	1	5	1	3.0	5	3	4
249/803/80	0	0	1	1	5	1	3.0	10	5	7.5
249/804/80	0	0	1	1	5.5	1	3.2	7.5	5	6.2
258/810/80	1	0	5	1	15	3	9.0	20	7.5	13.7
112/3200/74	trace	0	5	5	40	10	25	50	10.5	30.2
122/65/67	trace	0	10	5	35	3	19	40	5	22.5
138/1881/76	0	trace	10	5	40	10	25	35	10	22.5
133/2067/77	1	0	10	5	30	5	17.5	35	10	22.5
80-B/2528/70	0	0	15	7.5	22.5	10	16.2	30	15	22.5
90/1966/72	0	0	20	10	25	5.5	15.2	45	12.5	28.7
Protein 10/78	0	0	5	3	32.5	3	17.2	40	5	22.5
" 56/78	trace	trace	10	10	45	25	35.0	45	20	32.5
" 88/78	1	1	5	1	22.5	3	12.7	20	5	12.5
" 114/78	0	0	5	1	32.5	5.5	19.0	40	12.5	26.2
Comp. 2 S_2 L.29	0	0	10	5	32.5	7.5	20	35	15	25
" " L.32	0	0	5	1	20	3	11.5	25	7.5	16.2
" " L.58	0	0	12.5	1	32.5	3	17.7	35	5	20
H.B.P. 3 S_1 L.18	0	0	7.5	1	40	5	22.5	40	5	22.5
H.B.P. 6 S_3 L.1	0	0	15	1	22.5	5.5	14	32.5	10	21.2
H.B.P12 S_1 L.80	0	0	1	1	5	3	4	12.5	7.5	10
B.W.C./503/80	0	0	7.5	5	32.5	10	21.2	25	15	20
B.W.C./523/80-A	0	0	3	1	7.5	1	4.2	7.5	5	6.2
R.C/39/80	0	0	5	1	7.5	3	5.2	15	5	10
78 S 49456	0	0	5	1	15	1	8	25	3	14
Seville Giant	0	0	1	1	10	3	6.5	15	5	10
NEB/2727/75	0	0	7.5	1	30	5	17.5	35	5	20
ILB 938	0	0	1	1	10	1	5.5	10	5	7.5
Mean	0	0	7.5	3.4	23.3	5.9	14.5	27.5	9.1	18.3
L.S.D. 5%	-	-	-	-	15.4	8.0		6.2		9.3

NS - not sprayed   S - sprayed with Diathane M45

Table 4. Average % of infection of faba bean entries with rust, during the growing season when sprayed or not with Dithane M 45, Sakha 1980/81.

1980/81 season. The average percentage of infection of tested lines with rust disease is presented in Table 4. The studied genotypes significantly differed in their reaction to rust disease. Lines 249/801/80, 249/802/80, 249/803/80, 249/804/80, Seville Giant, R.C. 39/80 and ILB 938 showed the lowest percentage of infection. Dithane M 45 fungicide significantly decreased rust disease infection. Susceptible lines responded better to fungicide application.

1981/82 season. The average reaction of rust (Fig. 6) revealed significant difference at final estimation at Nubaria (16 March) and Sakha (15 April). Over the two locations lines 249/802/80, 249/803/80, 249/804/80 and ILB 938 were significantly lower than the check cultivar Giza 3. Over all the data Agrimycin application was not significant, with no significant interaction between genotypes and sprayed treatments. On the other hand, the fungicidal chemical control with Agrimycin decreased rust disease infection with 14.4%.

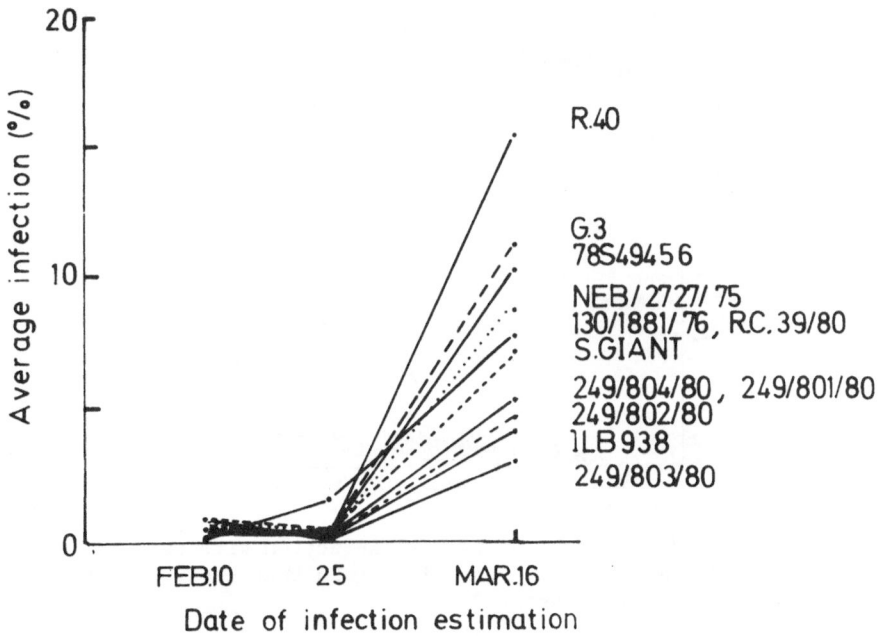

Figure 6. Development of field infection with rust on twelve faba bean genotypes at Nubaria.

1982/83 season. The incidence and development of rust infection on the tested genotypes was also estimated under artificial and natural epiphytotics. Inoculation with chocolate spot disease significantly decreased rust infection on the average of all tested genotypes at Nubaria with no significant genotype x inoculation interaction. However, at Sakha there was significant interaction as two lines, i.e. 249/803/80 similarly reacted to rust infection under both inoculation treatments, where as the other six lines showed significant increase under artificial inoculation

at the final estimate. Mayhew and Ford (1971) pointed out that <u>Physarum</u>
<u>polycephalum</u>, a mexomycete, produced an inhibitor which inhibited
infection of tobacco and bean by TMV and of <u>Vigna</u> <u>sinensis</u> by TRSV.
Environmental factors (temperature, humidity, etc.) and genetic
constitution of the host and pathogen might have an impact on the presence
of genotype x inoculation interaction at Sakha.

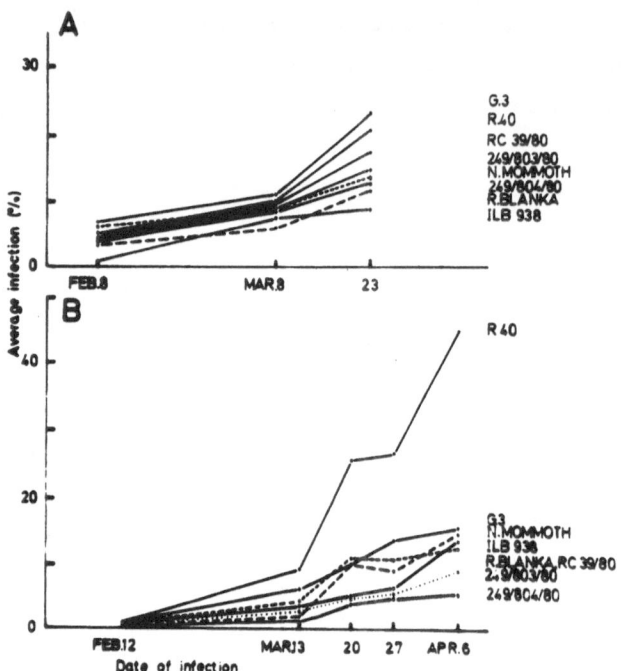

Figure 7. Development of natural infection with rust (average)
mean of the experiment) on eight faba bean genotypes at
Nubaria (A) and Sakha (B).

On the average of both inoculation treatments, reaction of genotypes
showed significant differences among tested lines on 8 February, 8 March
and 23 March at Nubaria (Fig. 7-A) and 13, 20, 27 March and 6 April at
Sakha (Fig. 7-B). Line ILB 938 and Reina Blanka at Nubaria and 249/804/80
at Sakha (Table 5) were resistant (slow rusting). ILB 938 and 249/804/80
were recorded as resistant at both locations during 1981/82 season.
However, both did not exhibit genotype x location interaction. Results of
Conner and Bernier (1982) indicated that faba bean rust isolates could be
divided into four races based on the reaction induced on faba bean inbreds
and pea cultivars.

Genotypes	Chocolate spot												Rust						
	Nubaria			Sakha					Giza				Nubaria			Sakha			
	2/8	3/8	3/23	2/12	3/13	3/20	3/27	4/6	2/7	2/12	2/18	3/15	2/8	3/8	3/23	3/13	3/20	3/27	4/6
Giza 3	7	7	7	3	7	7	7	7	4	7	7	7	7	8	8	7	6	7	7
Rebaya 40	8	8	8	7	8	8	8	8	8	8	8	8	5	7	7	8	8	8	8
ILB 938	1	1	1	1	1	1	2	3	1	1	1	1	1	2	1	5	2	3	5
R. Blanka	6	6	5	8	6	6	6	6	6	5	6	6	2	1	2	6	7	6	4
N. Mommoth	2	5	4	6	5	5	5	5	2	4	4	4	3	6	4	3	5	5	6
249/803/80	5	4	6	2	4	4	1	4	7	6	5	5	6	5	5	4	4	1	2
249/804/80	3	3	3	5	3	3	4	2	5	3	3	3	4	3	3	1	1	2	1
R.C. 39/80	4	2	2	4	2	2	3	1	3	2	2	2	8	4	6	2	3	4	3

Table 5.   Ranking order of eight faba bean genotypes for disease reaction at Nubaria, Sakha and Giza Research Stations, 1982/83.

It should be concluded that one line : ILB 938 was the most promising line for resistance to chocolate spot disease followed by two lines : 249/804/80 and R.C. 39/80.   Lines ILB 938, R. Blanka, 249/803/80 and 249/804/80 were resistant to rust.

REFERENCES

Bernier, C.C. 1983. Strategies for disease control in faba beans. A workshop of improvement : Faba beans, Kabuli chickpeas, and lentils, held at Aleppo, (ICARDA) 16-20 May, 1983.

Conner, R.L. and C.C. Bernier 1982. Race identification in Uromyces Viciae fabae. Can. J. Plant Path. 4, 157-160.

Conner, R.L. and C.C. Bernier 1982b. Slow rusting resistance in Vicia faba. Can. J. Plant Path. 4. 263-265.

El-Sherbeeny, M.H. and Mohammed, H.A. 1980. Detached leaf technique for infection of faba bean plants (Vicia faba L.) with Botrytis fabae. FABIS Newsletter, 2, 44-45.

Jellis, G.J., Bond, D.A. and Old, J. 1982. Resistance to chocolate spot (Botrytis fabae) in ICARDA accessions of Vicia faba. FABIS Newsletter, 4, 53-54.

Khalil, S.A. and J.G. Harrison 1981. Methods of evaluating faba bean materials for chocolate spot. FABIS Newsletter, No. 3, 51-52.

Mayhew, D.E. and Ford, R.E. 1971. An inhibitor of tobacco mosaic virus produced by Physarum polycephalum. Phytopathology 61, 636–640. C.F. Horsfall and Cowling, R. (Eds.) Plant Disease, An Advanced Treatise. Vol. 1. (1977).

Parlevliet, J.E. and Van Ommeren, A. 1975. Partial resistance of barley to leaf rust, Puccinia hordei. II. Relationship between field trials, micro test plots and latent period. Euphytica 24, 293–303.

Wilcoxson, R.D., Skovmand, B. and Atif, A.H. 1975. Evaluation of wheat cultivars for ability to retard development of stem rust. Ann. App. Bio. 80, 275–281.

PHYTOALEXIN PRODUCTION BY <u>VICIA FABA</u> - A MODEL SYSTEM FOR
THE STUDY OF GENE EXPRESSION

J.W. Mansfield, Biological Sciences Department, Wye College, Kent.
Yvonne M. Barlow and A.E.A. Porter, Chemistry Department, The University
Stirling, FK9 4LA, Scotland

Phytoalexins may be described as **low molecular weight antimicrobial compounds that are both synthesized by and accumulate in plants which have been exposed to microorganisms** (Mansfield & Bailey, 1982). In general, members of plant families usually produce chemically similar types of phytoalexin, for example isoflavonoids from the Leguminosae and terpenoids from the Solanaceae and although one compound may predominate in a particular species most plants produce several closely related phytoalexins. <u>Vicia faba</u> provides a notable exception to these generalizations; like most other legumes it produces an isoflavonoid phytoalexin, in this case medicarpin, but the principal induced antimicrobial compounds are furanoacetylenic wyerone derivatives. Structures of the eight furanoacetylenes recognised as phytoalexins in <u>V. faba</u> are given in Table 1. Hydrohydroxyketowyerone is found only in low concentration and was previously described as PA 4 (Hargreaves <u>et al</u>., 1977).

Recent developments in high performance liquid chromatography have allowed the separation of dihydro derivatives from their unsaturated analogues for quantitative analysis. Porter <u>et al</u>. (1979) and Mansfield <u>et al</u>. (1980) used rather complicated programmed elution of reversed phase columns but Barlow (1982) recently developed isocratic systems which are more efficient and reproducible. An example of the separation achieved with an ODS ultrasphere column (25 x 0.46 cm) using the solvent 1% aqueous formic acid:acetonitrile (40:60 v/v) is given in Fig. 1.

Figure 1.   Separation of a mixture of wyerone derivatives by isocratic h.p.l.c., solvent 1% formic acid:acetonitrile (40:60) u.v. detection at 330 nm, flow rate 0.8 ml min^{-1}; 1)wyerone acid; 2)dihydrowyerone acid; 3)wyerol; 4)wyerone epoxide; 5)dihydrowyerol; 6)wyerone; 7)dihydrowyerone. The internal standard <u>n</u> butyl salicylate had an average retention of 38.7 min. (over page)

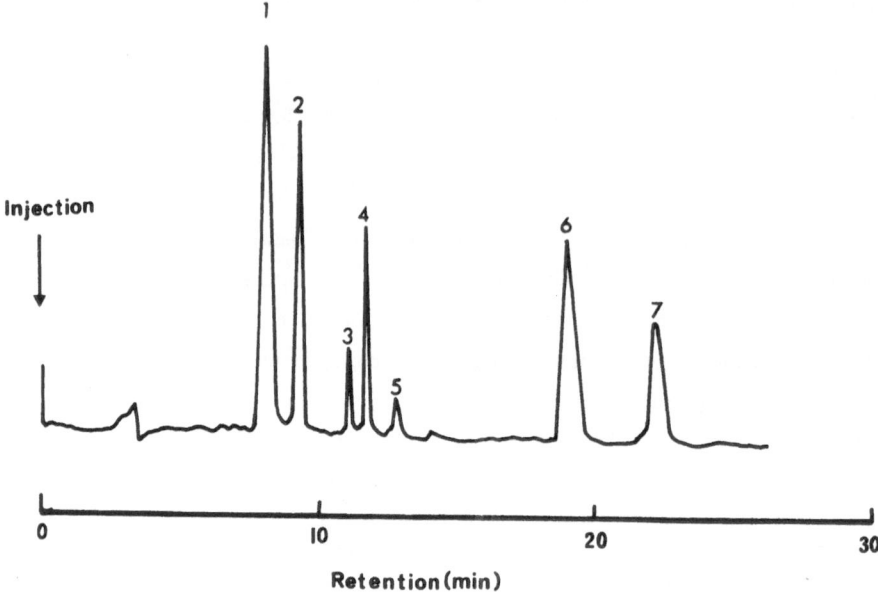

Injection

Retention (min)

Wyerone and dihydrowyerone are the predominant phytoalexins produced by cotyledons but wyerone acid and dihydrowyerone acid predominate in leaves and pods (Mansfield et al., 1980). Phytoalexin accumulation is particularly rapid in imbibed cotyledons challenged with Botrytis cinerea. Data presented in Fig. 2 show that the combined yields of wyerone, dihydrowyerone, wyerol, dihydrowyerol, wyerone epoxide, wyerone acid and dihydro wyerone acid reached 1290 ug g^{-1} fresh weight after 2 days. Wyerone alone reached concentrations >700 ug g^{-1}. The accumulation of these compounds represents a tremendous synthetic commitment by affected cells at inoculation sites.

Table 1. Structures of furanoacetylenic phytoalexins from Vicia faba (following page)

Me.CH$_2$.CH = CH.C ≡ C.CO—[furan]—CH = CH.COOR

1. Wyerone, R = Me
2. Wyerone acid, R = H

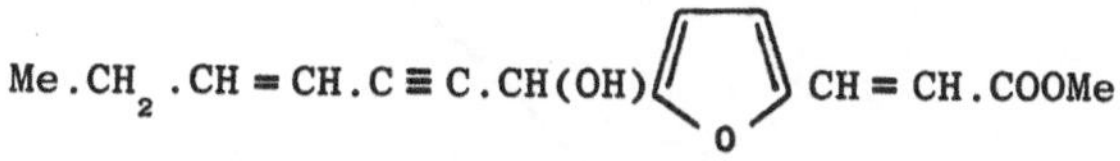

Me.CH$_2$.CH = CH.C ≡ C.CH(OH)—[furan]—CH = CH.COOMe

3. Wyerol

MeCH$_2$.CH$_2$.CH$_2$.C ≡ C.CO—[furan]—CH = CH.COOR

4. Dihydrowyerone, R = Me
5. Dihydrowyerone acid, R = H

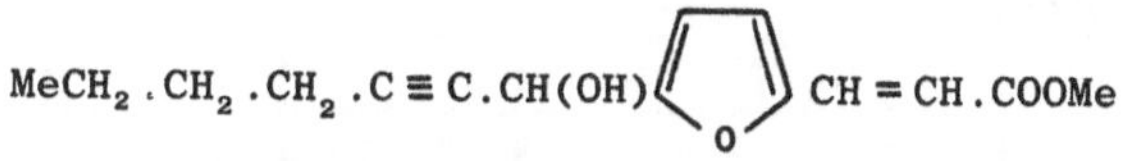

MeCH$_2$.CH$_2$.CH$_2$.C ≡ C.CH(OH)—[furan]—CH = CH.COOMe

6. Dihydrowyerol

Me.CH$_2$.CH —[O]— CH.C ≡ C.CO—[furan]—CH = CH.COOMe

7. Wyerone epoxide

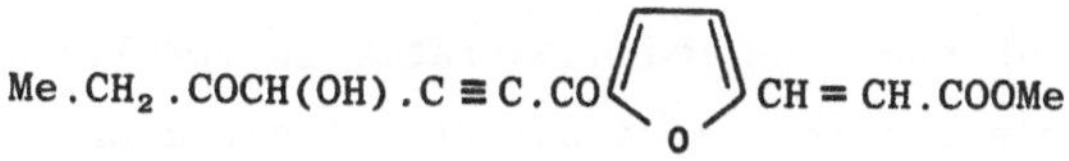

Me.CH$_2$.COCH(OH).C ≡ C.CO—[furan]—CH = CH.COOMe

8. Hydrohydroxyketowyerone

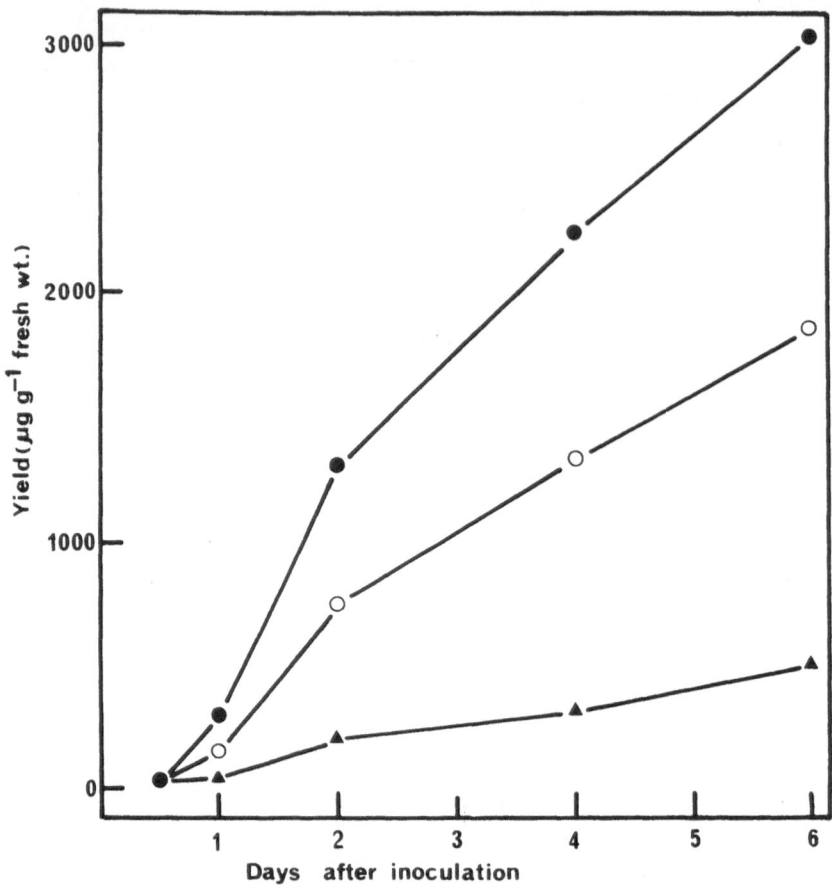

Figure 2. Changes in the concentrations of wyerone derivatives;   ,
1-7 combined (Table 1);   , wyerone and   , dihydrowyerone in
cotyledons inoculated with <u>Botrytis</u> <u>cinerea</u> (5 x 10^5
conidia ml^{-1}).

An important feature of the definition of phytoalexins given above is
that it restricts phytoalexins to compounds synthesized from remote
precursors. This has been confirmed for wyerone using radioactively
labelled precursors $[1-^{14}C]$ acetate $[2-^{14}C]$ malonate and $[\underline{n}\ 9,10-^3H]$
oleate. Application of the labelled substrates to cotyledons 15 h after
inoculation with <u>B.</u> <u>cinerea</u> gave incorporation figures of 0.16, 0.45 and
0.70% respectively five days after inoculation, clearly demonstrating the
induction of <u>de</u> <u>novo</u> wyerone biosynthesis (Cain <u>et</u> <u>al</u>., 1979). The trend
in incorporations, increasing from acetate through malonate to oleate is
consistent with an acetate plus <u>n</u> malonate pathway leading to oleate
during the early stages of wyerone biosynthesis. Details of the
necessarily complex sequence from oleate to wyerone are still mostly
unknown. Time course studies of the accumulation of wyerone and its

cogeners indicate that the furanoacetylemic derivatives may be synthesized in the following sequence:

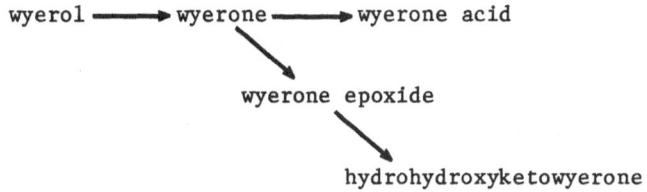

wyerol ⟶ wyerone ⟶ wyerone acid

wyerone epoxide

hydrohydroxyketowyerone

A similar sequence may be applied to the dihydro derivatives, i.e.

dihydrowyerol ⟶ dihydrowyerone ⟶ dihydrowyerone acid

The proportion of the dihydro derivatives to their unsaturated forms decreases with time after inoculation, this may be explained by the biosynthetic steps involving dihydro forms occurring at slower rates than those of their unsaturated analogues (Mansfield et al., 1980).

As a result of work with phaseollin, a phytoalexin from Phaseolus vulgaris, Hargreaves & Bailey (1978) and Bailey (1982) have proposed that phytoalexin biosynthesis is activated by endogenous elicitors released from cells dying as a result of exposure to microorganisms or various phytotoxic materials. It is envisaged that the endogenous elicitors diffuse from dying or dead cells into surrounding live cells which are induced to synthesize phytoalexins. If it is assumed that endogenous elicitors are the only compounds to have a direct effect on phytoalexin biosynthesis the following factors may be considered to control phytoalexin accumulation during lesion formation. 1) The amount of endogenous elicitor released from dead cells, 2) the time allowed for active synthesis to take place in living cells before they are overtaken by necrosis. If we make the further assumption that all dead cells release the same amounts of elicitor then we are led to the simple hypothesis (A) that treatments causing the same rates of necrotic lesion expansion will cause the same accumulation of phytoalexins. In order to test this hypothesis Barlow (1982) examined the relationship between accumulation of wyerone derivatives and cell death in the epidermis of leaves of V. faba following their treatment with B. cinerea and B. fabae, solutions of $AgNO_3$, $HgCl_2$, the surfactant Triton X-100 and a glucan isolated from cell walls of Phytophthora megasperma (PMS) which is known to cause phytoalexin accumulation in soybean (Ayres et al., 1976).

All treatments were applied to the abaxial epidermis of detached leaves. Droplets (20 µl) of spore suspension each contained c. 20 conidia. Both B. cinerea and B. fabae typically produce limited lesions from such a low level of inoculum. Solutions of the other compound were applied in 20 µl droplets to sites pricked with ten sterile fine steel pins. Preliminary experiments showed that wounding with pin pricks was required for $HgCl_2$ and $AgNO_3$ solutions to penetrate the cuticle and cause well defined areas of cell death. Epidermal strips were recovered from inoculation sites at intervals after inoculation and extracted for measurement of phytoalexin concentrations or examined microscopically following staining with neutral red (0.1% in 1 M sucrose) to determine the areas of epidermis killed around pin pricks or fungal penetration points.

All treatments caused some cell death and could be broadly separated into those causing large and small necrotic lesions roughly corresponding to the areas of cells killed by B. fabae and B. cinerea respectively. Thus PMS glucan, Triton X-100, and low concentrations of $HgCl_2$ and $AgNO_3$ caused areas of necrosis similar to that recorded for B. cinerea, other treatments were more comparable with B. fabae. The most toxic compound was 10^{-3} M $AgNO_3$ and dead areas around pin-pricks frequently coalesced 48 h after addition of the silver salt (see Fig. 3).

Pin-pricking with the addition of distilled water alone did not cause phytoalexin accumulation and only trace amounts were found in epidermis treated with Triton X-100. Quantifiable levels of wyerone derivatives were recovered following all other treatments: yields recovered are expressed per unit area of dead cells in Fig. 3. As proportions of each wyerone derivative were usually similar irrespective of treatment applied only data for wyerone acid (the predominant phytoalexin) and total wyerone derivatives have been present.

Chemical treatments causing little necrosis caused only low levels of phytoalexin accumulation. By contrast, B. cinerea and the PMS glucan, in spite of causing equally small lesions, elicited comparatively high levels of accumulation after 24 and 48h. The increase in yields between 24 and 48 h was quite striking following treatments with PMS glucan. All of the treatments producing large lesions (>0.6 mm^2 after 24 h) caused rapid phytoalexin accumulation.

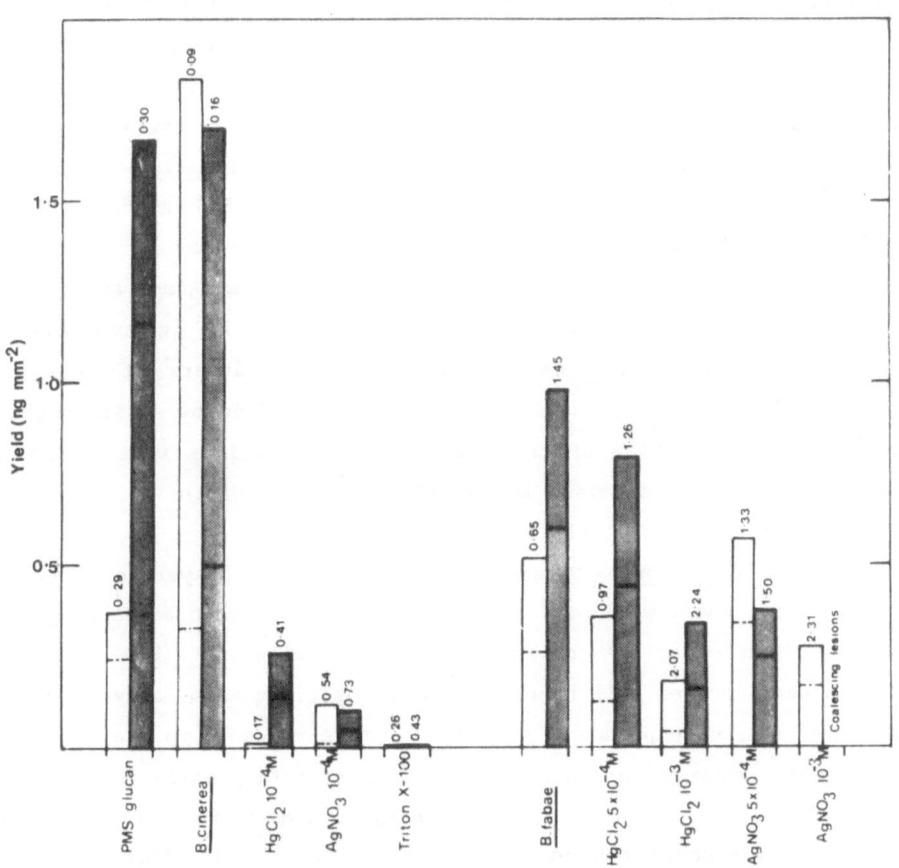

Figure 3. Combined yields of wyerone derivatives 1-7 (Table 1)
based on the area of the zone of cells killed around pin-
pricks or penetration points 24h (open columns) and 48h
(shaded columns) after treatments causing small or large
lesions. Lesion size (mm^2) is given above each histogram.
The broken line indicates the yield of wyerone acid.

Differences observed in phytoalexin accumulation following treatment
with **B. cinerea** or PMS glucan and Triton X-100 or heavy metals clearly
demonstrate that hypothesis (A) is untenable. Where small lesions are
produced treatments causing the same rates of cell death do not cause the
same accumulation of wyerone derivatives. However, the hypothesis has not
been disproved for treatments causing larger lesions. Differences
observed between phytoalexin accumulation in response to 10^{-3} and 5×10^{-4}
M $HgCl_2$ or 10^{-3} and 5×10^{-4} M $AgNO_3$ can be simply explained in terms of
rates of lesion expansion. Very rapid lesion expansion (as with 10^{-3} M
concentrations) may allow release of endogenous elicitors but limit time

available for biosynthesis. More gradual expansion of lesions (as with 5 x 10^{-4} M concentrations of B. fabae) caused greater phytoalexin accumulation and may therefore provide not only continued release of endogenous elicitors but also sufficient time for phytoalexin accumulation through biosynthesis in live cells surrounding necrotic tissue.

The greater efficiency of B. cinerea and PMS glucan eliciting phytoalexin accumulation without killing many cells may be explained in two ways. 1) The treatments may kill cells in such a manner as to release large amounts of endogenous elicitors. 2) The PMS glucan and some component of B. cinerea may act directly as inducers of phytoalexin synthesis. These two possibilities could be examined by an investigation of the modes of action of the glucan and the elicitor from B. cinerea. Such investigations may lead to the identification of the molecules which derepress genes for phytoalexin biosynthesis.

The rapid de novo synthesis and accumulation of wyerone derivatives is now sufficiently well defined to allow the process to be developed as a model for studies on the control of gene expression in V. faba. The first step in such studies will be to examine the activities of enzymes involved in the biosynthetic pathway. It should prove valuable to examine enzymes involved in later conversions as well as those acting early in the pathway for example the enzymes of fatty acid biosynthesis. The mechanisms leading to increased and novel enzyme activities could then be examined using comparative density labelling as applied by Lamb et al. (1980) to demonstrate de novo synthesis of phenylalanine ammonia lyase during accumulation of the phytoalexin phaseollin in P. vulgaris. The application of other sensitive techniques such as immunoprecipitation of enzymes and measurement of mRNA synthesis using reticulocyte translation systems should quickly extend our knowledge of protein synthesis de novo and gene function in Vicia. Clearly we are some way from achieving these goals but the phytoalexin accumulation model appears to have sufficient potential to merit commitment to such a research programme.

REFERENCES

Ayers, A.R., Ebel, J., Valent, B. & Albersheim, P. 1976. Host pathogen interactions X. The fractionation and biological activity of an elicitor isolated from the mycelial walls of Phytopthora megasperma var. sojae. Plant Physiology, 57, 760-765.
Bailey, J.A. 1982. Mechanisms of phytoalexin accumulation. In: Phytoalexins eds. J.A. Bailey and J.W. Mansfield, Blackie & Sons, Glasgow, pp. 289-318.

Barlow, Y.M. 1982. The accumulation of phytoalexins in Vicia faba L. and their analysis by high performance liquid chromatography. M.Sc. thesis, University of Stirling.

Cain, R.O. & Porter, A.E.A. 1979. Biosynthesis of the phytoalexin wyerone in Vicia faba. Phytochemistry, 18, 322-323.

Hargreaves, J.A. & Bailey, J.A. 1978. Phytoalexin production by hypocotyls of Phaseolus vulgaris in response to constitutive metabolites released by damaged cells. Physiological Plant Pathology, 13 89-100.

Hargreaves, J.A., Mansfield, J.W. & Rossall, S. 1977. Changes in phytoalexin concentrations in tissues of the broad bean plant (Vicia faba L.) following inoculation with species of Botrytis. Physiological Plant Pathology, 11, 227-249.

Lamb, C.J., Lawton, M.A., Taylor, S.J. & Dixon, R.A. 1980. Elicitor modulation of phenyl ammonia-lyase in Phaseolus vulgaris. Annales Phytopathologie, 12, 422-433.

Mansfield, J.W. & Bailey, J.A. 1982. Phytoalexins: current problems and future prospects, in Phytoalexins, eds. J.A. Bailey and J.W. Mansfield, Blackie & Sons, Glasgow, pp. 319-323.

Mansfield, J.W., Porter, A.E.A. & Smallman, R.V. 1980. Dihydrowyerone derivatives as components of the furanoacetylenic phytoalexin response of tissues of Vicia faba. Phytochemistry 19, 1057-1061.

Porter, A.E.A., Smallman, R.V. & Mansfield, J.W. 1979. Analysis of furanoacetylenic phytoalexins from the broad bean plant by high performance liquid chromatography. Journal of Chromatography, 172, 498-504.

A NOTE ON DISEASE INTERACTION IN HOST PLANTS OF VICIA FABA

Said M. Omar

I.C.A.R.D.A., Egypt

In many parts of the world the number of different pathogens attacking field bean can be very large. The development of a disease on its host, sometimes can be greatly affected by the presence of another. In the present work conducted at Wye College, interaction between viral and fungal diseases of Vicia faba L. were investigated. Bean yellow mosaic virus, or Bean leaf roll virus stimulated development of chocolate spot fungus. The results were confirmed in laboratory, glasshouse and field trials. Increased susceptibility to the fungus was positively correlated with virus replication in the host tissues. Increased nutrient substances (carbohydrate and amino acids) in the diffusate of virus infected leaves possibly stimulated Botrytis growth and consequently more damage occurred.

A synergistic reaction was also detected in the root of BYMV-infected plants additionally infected with Rhizoctonia solani. Increased electrolyte leakage from virus-infected plants over healthy ones was shown to be a likely cause of increased root rot infection.

In contrast, acquired resistance to rust fungus was induced by preinoculation of the host plant with BLRV or BYMV. This protection was more pronounced when the virus had been established in the host for about two weeks.

Conversely, virus infection was not adversely affected by prior inoculation with rust and the rust infection had a negligible effect on virus infectivity.

The impact of virus infection on growth and yield was significantly greater than that caused by either rust or chocolate spot. It was found that combinations of fungus and virus caused greater yield losses and reference to single and combined infections of the various pathogens gave indications of a possible stepwise effect on yield loss. Finally, if Botrytis resistance were claimed by a breeder, it would in the light of the work reported here, be realistic to have resistance tested partially in plants debilitated by previous virus inoculation.

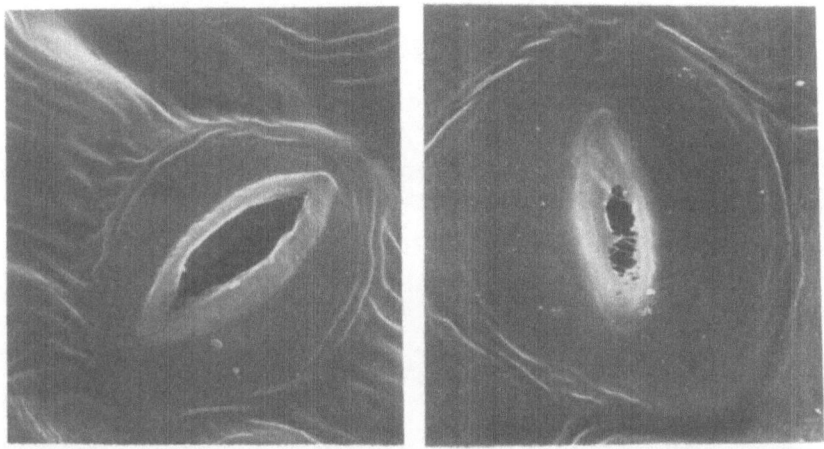

Figure 1.   Scanning electron microscope images showing variation
in stomatal type in a) healthy and b) BYMV infected leaves.
Bar = 5.0 µm.

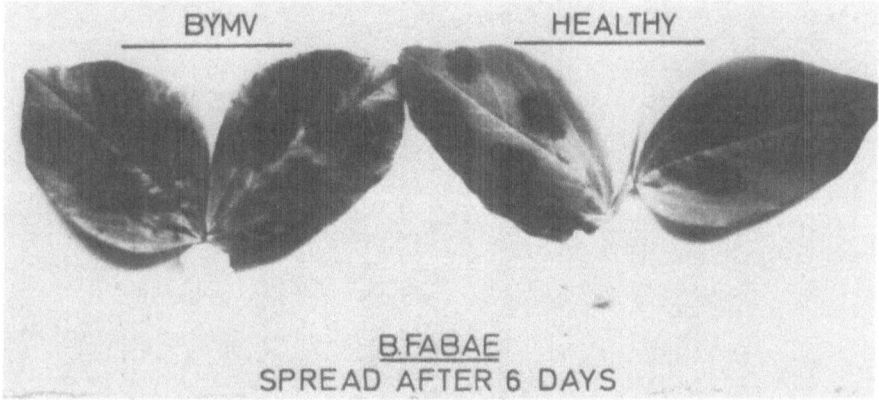

Figure 2.   Detached leaf test showing the effect of BYMV infection
on the development of lesions caused by B. fabae.

4
Population

THE USE OF ISOZYME GENES AS MARKERS IN THE
POPULATION GENETICS OF VICIA FABA L.

W. E. Peat[1] and J. Y. Adham[2].

[1]Wye College (University of London), Ashford, Kent, U.K.

[2]ICARDA, P.O. Box 5466, Aleppo, Syria.

ABSTRACT

The extraction and identification of 26 esterase isozyme bands from newly expanded leaves of Vicia faba is described. Eight of these bands were shown to be controlled by three genes: one (designated A) with two alleles and two (B & C) each with three alleles. An experiment is reported showing the use of these isozyme gene to estimate outcrossing frequencies in bean populations enclosed in pollination cages. The use of isozyme techniques in other aspects of the population genetics of Vicia faba is discussed.

INTRODUCTION

The use of isozymes as marker genes has become a common technique in many areas of plant genetics. In Vica faba, variation has been reported in a wide range of enzyme systems and has been used in evolutionary and taxonomic studies (Ladizinsky, 1975; Yamamoto, 1975, 1979; Yamamoto & Plitmann, 1980), inbred line recognition (Gates & Boulter, 1979, 1980) and cultivar identification (Bassiri & Rouhani, 1977; Kaser & Steiner, 1983). Most of these papers report genetic variation but have not given any information of the genetic control of the observed banding patterns.

The segregation of a single gene controlling a dimeric form of glutamate-oxaloacetate-transaminase was reported by Suso and Moreno (1982), and preliminary data on inbred parent and F_1 generations for several enzymes by Gates & Boulter (1979, 1980) and De Mora, Gonzalez & Serradilla (1983). With the exception of one report on pollen isozymes (Gates & Boulter, 1980), all of this work was based on seed extracts.

Seeds are useful sources of material since they provide stable, consistent samples which can be stored and chemically analysed at leisure. Analysis is, however, destructive and this complicates the genetic interpretation. For example, F_2 genotypes need to be inferred from a sample of their F_3 progeny, and it is not possible to relate an individual's genotype directly with its performance. The use of pollen isozymes avoids many of these problems, provided that the analyses are sufficiently sensitive to detect the very small quantities of enzyme available in small amounts of pollen.

The extraction of enzymes from vegetative tissues (especially leaves) removes some of these difficulties. The quantity of tissue (and therefore of enzyme) is rarely limiting, and non-destructive sampling is possible. The main disadvantage is that many enzymes of leaf tissue can show variation due to environmental or tissue-ageing effects which can complicate (or even prevent) genetic interpretation. Provided enzymes can be identified which do not show these environmental effects, then the use of leaf tissue has advantages over all other options.

In this paper we report evidence on the genetic control of three genes determining non-specific esterase enzymes extracted from young leaf tissue and demonstrate their application to pollination studies. Most of the work was carried out at ICARDA, and we gratefully acknowledge the assistance and support of the Director General and of Dr. G. Hawtin.

MATERIALS AND METHODS

The topmost expanded leaf was collected from plants at the 4 - 6 leaf stage, quickly frozen and stored until assayed. Approximately 0.5g tissue was crushed in 2ml of an extractant containing 20% w/v sucrose and 1% bromophenol blue as a marker dye. Frozen extracts could be stored for several weeks if necessary with no loss of activity. Electrophoresis was performed through 7% polyacrylamide gels which were then assayed for non-specific esterases by their ability to hydrolyse naphthyl acetate combined with staining with Fast Blue R.R. salt (Brewbaker, Upadhaya, Makinen and MacDonald, 1968).

To determine the inheritance of particular band patterns, individual plants were assayed, controlled pollinations carried out and their progeny examined. A range of self- and cross-pollinations were performed to provide F_1, F_2 and Backcross material suitable for assessing Mendelian segregation ratios.

The use of isozyme variants as markers for pollination studies was tested in an experiment performed at Aleppo between November 1980 and June 1981, using 7 x 3.5m pollination cages. Each cage was sown with the Egyptian variety Giza 4 into which were dispersed 12 plants of the Syrian variety ILB 1812 and 12 plants of Triple White. ILB 1812 was chosen since it possessed different frequencies of the bands for which a genetic interpretation had been made, and Triple White was used to provide an independent test of cross-pollination, using the hilum-colour gene as a marker. The experiment was designed with four treatments: two planting densities of 8 and 16 plants m^{-2}, combined with two bee populations of one or two hives per cage during the flowering period. Each treatment was replicated four times. During the growing season, each plant of ILB 1812 and a sample of plants of Giza 4 from each cage were assayed for their esterase pattern. At the end of the season, seed was collected from each plant of ILB 1812 and from each plant of Triple White. These progenies were grown and assayed for their esterase pattern or hilum colour respectively. Outcrossing frequencies were determined using the methods of Brown and Allard (1970) for isozyme genes and of Fyfe and Bailey (1951) for hilum colour.

RESULTS

The electrophoresis of esterase isozymes from newly-expanded leaves of plants from a wide range of cultivars revealed a total of 26 bands. Eight of these gave patterns which were sufficiently stable to permit genetic interpretation. These are the regions labelled A, B & C in Fig. 1.

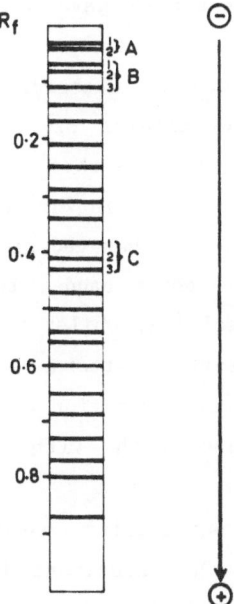

Fig. 1   The 26 esterase bands extracted from <u>Vicia faba</u> leaves, drawn in their positions relative to the bromophenol blue marker dye.

TABLE 1.   Progeny of controlled self- and cross-pollinations of
plants assayed for the presumed gene A.

Parent Band Pattern	No. of Families	Offspring band Pattern			Expected Ratio		
		A_1A_1	A_1A_2	A_2A_2			
A_1A_1 selfed	5	125	0	0	1	-	-
A_1A_1 * A_2A_2	3	0	8	0	-	1	-
A_2A_2 selfed	6	0	0	161	-	-	1
A_1A_2 selfed	9	79	167	77	1	2	1
A_1A_1 * A_1A_2	4	12	9	0	1	1	-
A_2A_2 * A_1A_2	2	0	8	6	-	1	1

All the plants examined possessed either one densely staining band or
both bands more lightly stained at the positions A_1 and A_2, with R_f's of
0.03 & 0.04 respectively. Test crosses showed that single-banded
individuals always bred true on selfing, whilst crossing the different
single-banded types produced the double-banded offspring. Selfing double-
banded individuals produced a 1:2:1 ratio, whilst crossing single-banded
individuals with the double-banded gave 1:1 ratios (Table 1). In all
crosses, the observed segregations were in agreement with Mendelian
expectation. It may be concluded, therefore, that these two bands are
produced by two co-dominant alleles of a single gene. There was no
apparent linkage between any of these three genes, but the number and size
of families providing evidence for this was small.

The regions labelled B & C behaved similarly to each other: plants
possessed one or two of the three bands, but never all three. Single-
banded plants bred true on selfing, whilst two-banded plants showed 1:2:1
ratios. Crosses between assorted patterns also gave segregations which
were in agreement with the interpretation that both of these regions were
controlled by single genes each with three co-dominant alleles
(Tables 2 and 3).

The determination of outcrossing frequency is illustrated for a
single pollination cage in the experiment (Table 4). Genotypes of the
female parents were known, and the frequencies of alleles in the pollen
population had also been estimated (Table 5). In Table 4, the expected
frequencies of each progeny genotype are shown, as given by Brown & Allard
(1970), with p and q being the allelic frequencies for A_1 and A_2
respectively, and t the outcrossing frequency. Since p and q were known,

TABLE 2   Progeny of controlled self- and cross-pollinations of plants assayed for the presumed gene B.

Parent Band Pattern	No. of Families	Offspring band Pattern						Expected Ratio
		B_1B_1	B_1B_2	B_1B_3	B_2B_2	B_2B_3	B_3B_3	
B_1B_1 selfed	2	34	0	0	0	0	0	1 - - - - -
B_2B_2 selfed	2	0	0	0	53	0	0	- - - 1 - -
B_3B_3 selfed	1	0	0	0	0	0	9	- - - - - 1
B_1B_2 selfed	5	46	93	0	55	0	0	1 2 - 1 - -
B_1B_3 selfed	4	27	0	42	0	0	19	1 - 2 - - 1
B_2B_3 selfed	6	0	0	0	43	91	37	- - - 1 2 1
B_1B_1 * B_1B_2	3	6	8	0	0	0	0	1 1 - - - -
B_1B_2 * B_1B_3	2	2	2	1	0	2	0	1 1 1 - 1 -
B_1B_2 * B_2B_3	2	0	1	2	2	2	0	- 1 1 1 1 -
B_1B_2 * B_3B_3	1	0	0	3	0	4	0	- - 1 - 1 -
B_1B_3 * B_2B_2	4	0	10	0	0	13	0	- 1 - - 1 -
B_1B_3 * B_2B_3	2	0	3	4	3	3	0	- 1 1 1 1 -
B_2B_3 * B_3B_3	3	6	8	0	0	0	0	1 1 - - - -

TABLE 3   Progeny of controlled self- and cross-pollinations of plants assayed for the presumed gene C.

Parent Band Pattern	No. of Families	Offspring band Pattern						Expected Ratio
		C_1C_1	C_1C_2	C_1C_3	C_2C_2	C_2C_3	C_3C_3	
C_3C_3 selfed	1	0	0	0	0	0	117	- - - - - 1
C_2C_2 * C_3C_3	4	0	0	0	0	21	0	- - - - 1 -
C_1C_2 selfed	1	6	16	0	5	0	0	1 2 - 1 - -
C_1C_3 selfed	4	69	0	130	0	0	74	1 - 2 - - 1
C_2C_3 selfed	3	0	0	0	46	77	45	- - - 1 2 1
C_1C_1 * C_2C_3	2	0	6	5	0	0	0	- 1 1 - - -
C_1C_2 * C_3C_3	3	0	0	7	0	5	0	- - 1 - 1 -
C_1C_3 * C_2C_2	5	0	15	0	0	12	0	- 1 - - 1 -
C_1C_3 * C_3C_2	2	0	0	6	0	0	5	- - 1 - - 1

each row in the table provided a separate estimate of t. These were combined into a weighted mean for the whole plot, using the size of each progeny as its weight. Genes B & C, each with three alleles, could similarly be used to estimate t. With these genes there were three allelic frequencies to be estimated, p, q & r; but once these were known the calculation of t was no more difficult. Since the same progenies were used to estimate t for all three genes, the estimates cannot be regarded as independent. The three genes were, however, segregating independently, so that the values obtained provided checks on the accuracy of estimation.

TABLE 4   The determination of cross-pollination frequency in plants of ILB 1812 surrounded by Giza 4: results of one plot.

Parent Genotype	No. of Families		Progeny Genotype		
			A_1A_1	A_1A_2	A_2A_2
A_1A_1	5	obs. no.	28	17	0
		obs. freq.	0.622	0.378	
		exp. freq.	$(1 - qt)$	qt	–
A_1A_2	2	obs. no.	2	9	7
		obs. freq.	0.111	0.50	0.389
		exp. freq.	$(1-t+2pt)/4$	0.5	$(1-t+2qt)/4$
A_2A_2	5	obs. no.	0	1	44
		obs. freq.	0.0	0.022	0.978
		exp. freq.	–	pt	$(1-pt)$

TABLE 5. Allelic frequencies of esterase genes A, B & C in plants of Giza 4 sampled as pollen parents

	Allele (band number)		
Gene	1	2	3
A	0.07	0.93	–
B	0.76	0.03	0.21
C	0.08	0.04	0.88

All three isozyme genes and the hilum colour gene showed similar treatment effects in the whole experiment. There was a highly significant difference in mean outcrossing frequency between the two planting densities, with the lower density showing a higher value (Table 6).

Conversely, there was no significant effect of bee population, probably since this was more than adequate in all cages. The estimates of t made from the three isozyme genes were in close agreement with each other, but were all larger than the estimates made using the hilum colour gene This was probably due to a low frequency of the white hilum allele in Giza 4, which was not allowed for in estimating t.

TABLE 6 Outcrossing frequencies determined from the segregation of three isozyme genes and (independently) from the segregation of hilum colour.

	16 plants m^{-2}		8 plants m^{-2}		
	1 Hive	2 Hives	1 Hive	2 Hives	s.e.
Gene A	0.403	0.322	0.528	0.540	
Gene B	0.435	0.348	0.525	0.537	0.041
Gene C	0.406	0.350	0.510	0.550	
White hilum	0.383	0.308	0.480	0.523	0.036

DISCUSSION

Isozyme variation has been discovered in so many plant species, and for so many different enzymes, that it must now be regarded as a normal part of the genetic variability of most plant populations. In contrast, modern agricultural species have been selected so rigorously for uniformity of appearance that few cultivars reveal much variation for genes with a visible effect. This means that population genetics using visible markers is only possible by introducing variant alleles into the population, and this is possible only under experimental conditions. Studies of the breeding system during the normal procedures of breeding and seed multiplication are impossible since the techniques of measurement would introduce undesirable contaminants. The use of isozymes as marker genes avoids these problems, and this ability more than compensates for the complexity of extraction and identification.

The estimation of outcrossing frequencies, as reported here, is one of the simpler applications of isozymes, and the theoretical background has been described in detail (Brown & Allard, 1970). Only a single gene is needed, and the experimental technique can be simpler than that described: provided relatively large progenies are collected from each female plant

(Brown & Allard, loc. cit. suggest 9 to 10 as "reasonably good"), female genotypes may be inferred from their progeny's segregation. Similarly, provided that female parents with two or more of the possible genotypes are used, allelic frequencies in the pollen pool and outcrossing frequency may be jointly estimated, using maximum likelihood methods. It is therefore possible to apply the technique retrospectively. This could, for example, provide useful information about the changes in heterozygosity accompanying seed multiplication in composite varieties, as well as providing checks on contamination in more highly inbred material (Gates & Boulter, 1979; 1980).

The range of isozyme variants which have been positively associated with known gene segregations in Vicia faba is so far small: apart from the three esterase genes reported here, there is only the report of Suso and Moreno (1982) on the segregation of a single gene controlling a dimeric form of glutamate-oxaloacetate-transaminase in seeds. Differences between inbred parent and F_1 hybrid plants has been demonstrated for several seed and pollen enzymes by Gates & Boulter (1979, 1980) and De Mora, Gonzalez & Serradilla (1983). Genetic analysis of subsequent generations derived from such material would be a useful addition to the available information.

The ability to monitor genetic change in populations makes it possible to relate isozyme genes to other phenotypic characters (Tanksley & Rick, 1980). It is necessary to characterise individuals for a large number of isozyme genes, distributed around the genome in known linkage relationships. Given such information, it is possible to relate the response to selection for economically important (polygenic) characters with frequency changes at known chromosomal regions (e.g. Stuber, Goodman & Moll, 1982 in maize), or alternatively to select for specific isozyme genes which are associated with economic characters (e.g. Stuber et al., 1980, also in maize). The application of such techniques to V. faba requires the extension of existing data by the discovery of more isozyme genes and their mapping and association (if possible) with particular chromosomes. This is recommended as a prority area of research in faba bean genetics.

REFERENCES

Bassiri, A. and Rouhani, I. (1977). Identification of broad bean cultivars based on isoenzyme patterns. Euphytica 26: 279 - 286.

Brewbaker, J. L., Upadhaya, M. D., Makinen, Y. and MacDonald, T. (1968). Isoenzyme polymorphism in flowering plants. III Gel electrophoresis methods. Physiol. Plant. 21: 930 - 940.

Brown, A. H. D. and Allard, R. W. (1970). Estimation of the mating system in open-pollinated maize populations using isozyme polymorphisms. Genetics 66: 133 - 145.

Fyfe, J. L. and Bailey, N. T. (1951). Plant breeding studies in leguminous forage crops. I. Natural cross-breeding in winter beans. J. agric. Sci., Cambs. 41: 371 - 378.

Gates, P. and Boulter, D. (1979). The use of seed isoenzymes as an aid to the breeding of field beans (Vicia faba L.). New Phytol. 83: 783 - 791.

Gates, P. and Boulter, D. (1980). The use of pollen isoenzymes as an aid to the breeding of field beans (Vicia faba L.). New Phytol. 84: 501 - 504.

Kaser, H. R. and Steiner, A. M. (1983). Subspecific classification of Vicia faba L. by protein and isozyme patterns. FABIS 7: 19 - 20.

Ladizinsky, G. (1975). Seed protein electrophoresis of the wild and cultivated species of the section faba of Vicia. Z. Pflanzenzuchtung 83: 785 - 788.

De Mora, T., Gonzalez, J. A. and Serradilla, J. M. (1983). Preliminary studies on genetic determination of electrophoresis alloenzyme bands in Vicia faba L. FABIS 7: 11 - 12.

Stuber, C. W., Goodman, M. M. and Moll, R. H. (1982). Improvement of yield and ear number resulting from selection at allozyme loci in a maize population. Crop Science 22: 737 - 740.

Stuber, C. W., Moll, R. H., Goodman, M. M., Schaffer, H. E. and Weir, B. S. (1980). Allozyme frequency changes associated with selection for increased grain yield in maize (Zea mays L.). Genetics 95: 225 - 236.

Suso, M.-J. and Moreno, M.-T. (1982). Genetic control of electrophoretic variation for glutamate-oxaloacetate-transaminase (G.O.T.) in Vicia faba L. FABIS 5: 14.

Tanksley, S. D. and Rick, C. M. (1980). Isozymic gene linkage map of the Tomato: applications in genetics and breeding. Theor. appl. Genet. 57: 161 - 170

Yamamoto, K. (1975). Estimation of genetic homogeneity by isozymes from interspecific hybrids of Vicia. I. Jpn. J. Breed. 25: 60 - 64.

Yamamoto, K. (1979). Estimation of genetic homogeneity by isozymes from interspecific hybrids of Vicia. II. Jpn. J. Breed. 29: 59 - 65.

Yamamoto, K. and Plitmann, U. (1980). Isozyme polymorphism in species of the genus Vicia (Leguminosae). Jpn. J. Genetics 55: 151 - 164.

STANDARDS EMPLOYED IN DISTINCTNESS, UNIFORMITY AND STABILITY TESTS OF FABA BEAN CULTIVARS

J. Higgins and J.L. Evans

National Institute of Agricultural Botany, Cambridge, U.K.

ABSTRACT

When a new faba bean cultivar is entered for a grant of Plant Breeders' Rights or for listing on the EEC Common Catalogue of cultivars it must be tested for distinctness, uniformity and stability (DUS). The standards operated in DUS tests in the U.K. are described.

INTRODUCTION

The end product of the faba bean breeding programme is the cultivar, where a cultivar is defined in the International Code of Nomenclature for Cultivated Plants as:

'An assemblage of cultivated plants which is clearly distinguished by any characters (morphological, physiological, cytological, chemical or others) and which, when reproduced (sexually or asexually), retains its distinguishing characters' [5].

The breeder may, if he wishes, apply for a grant of Plant Breeders Rights (PBR) in respect of his new cultivar. In the UK this is possible under the Plant Variety and Seeds Act 1964 which grants the holder of PBR, among other rights:

'....... the exclusive right to do, and to authorise others to do, as follows -
(a) to sell the reproductive material of the plant (cultivar)

(b) to produce the reproductive material of the plant (cultivar) in Great Britain for the purpose of selling it' [3].

Several other countries both inside and outside the EEC offer PBR for faba beans. In order to qualify for a grant of PBR the new cultivar must fulfil the requirements for distinctness and stability embodied in the above definition of a cultivar. In addition it must be uniform. This is laid down and stated in the 1964 Act as follows:

'The cultivar must be sufficiently uniform or homogenous having regard
to the particular features of its sexual reproduction or vegetative
propagation.' [3]

In the UK tests for distinctness, uniformity and stability (DUS tests)
on faba beans are carried out by the National Institute of Agricultural
Botany (NIAB), Cambridge, on behalf of the Ministry of Agriculture,
Fisheries and Food. A test is normally of two years' duration. The new
cultivar is grown in Spring-sown field plots together with any cultivars
known to be similar (controls). These are selected on the basis of
information about the new cultivar supplied by the breeder in a completed
Technical Questionnaire (TQ). In order to facilitate the selection of
appropriate controls all the cultivars known have been classified. A
simple artificial classification system is used — the initial groupings
being made on the basis of the most easily observed qualitative
characters. According to the International Union for the Protection of
New Varieties of Plants (UPOV), which is the inter-governmental
organisation of states interested in plant variety protection, qualitative
characters are:

'those which show discrete discontinuous states with no arbitrary
limit on the number of states.' [6]

The qualitative characters and their states used for classification are:

Character	States
Flower : melanin wing spot	absent/present
Seed : testa colour	grey-white/beige green/red/violet/ black
Seed : hilum colour	non-black/black

Within the resulting groups cultivars are further differentiated and
described using quantitative characters which, according to UPOV, are:

'those which are measured on a one dimensional scale and show continuous variation from one extreme to the other.' [6]

Examples of quantitative characters are:

Stem	: length	measure
Seed	: 100 seed weight	weight
Pod	: Number of ovules	count
Flower	: extent of standard petal anthocyanin	score

In the DUS test four plots of 30 plants are grown in a randomised block design. Records are taken of up to seven qualitative characters and up to 18 quantitative characters. A further two recorded characters are handled qualitatively but are in fact quantitative in nature. They are:

Seed	: shape	circular/elliptic/ square/oblong/ovate
Seed	: dimpling	absent/present

If a breeder wishes to have his new faba bean cultivar marketed within the EEC it must be added to the National List (NL) of faba bean cultivars of one of the member states. A DUS test is also required for this purpose and in the UK the test is exactly the same as for PBR except that the range of controls is narrower, being restricted to those cultivars listed or entered for listing within the EEC. As for PBR the test is carried out at NIAB, Cambridge. If a cultivar is entered for both PBR and NL one test covers both.

Once placed on the National List the name of the new variety then goes forward to be entered on the EEC Common Catalogue, which is a compendium of the National Lists of the individual member states.

STANDARDS

Distinctness

The variety must be 'clearly distinguishable', but what constitutes a clear difference? UPOV has attempted to define a clear difference for both qualitative and quantitative characters. Firstly the difference must be consistent, which in practice means repeatable. For true qualitative characters:

> 'The difference between two cultivars has to be considered clear if the respective characteristics show expressions which fall into two different states.' [6]

This means, for example, that a cultivar with melanin wing spots (one state) is distinct from one without melanin wing spots (another state). Similarly a cultivar with green testas (one state) is distinct from one with grey-white testas (another state).

The UPOV recommendation for quantitative characters which are handled qualitatively is that:

> 'an eventual fluctuation has to be taken into account in establishing distinctness.' [[6]

In other words, distinctness cannot necessarily be established simply because cultivars exhibit different character states, for the difference may in reality be very small and may not be easily observable in other growing seasons.

For the quantitative characters determination of distinctness may be possible simply by field plot observations as, for example, if the new cultivar is particularly early to flower or if there is a conspicuous height difference. Quite often, however, the differences may be small or sometimes there are no observable differences. It is then necessary to record a range of quantitative characters for the new cultivar and the controls. For these characters the UPOV recommendation is that:

> 'the difference has to be considered clear if it occurs with one per cent probability of an error, for example, on the basis of the method of the Least Significant Difference. The differences are consistent, if they occur with the same sign in two consecutive, or in two out of three, growing seasons.' [6]

Having 'the same sign' means that both significant differences must be in the same direction. This standard is applied at NIAB.

It may be that a clear difference between two cultivars cannot be found in any single character but that they may differ in several separately assessed characterists each of which on its own may not be statistically different (p ≤ 0.05). UPOV allows that such combinations of small differences can be held to satisfy the regulations but only if they are repeatable.

Uniformity

Faba beans are partially out-breeding and for this reason cultivars can exhibit high levels of variability, depending on the skill of the breeder. This is especially true for synthetic cultivars which are the result of inter-crossing several genetically different inbred lines. UPOV takes into account the inherent variability in faba bean cultivars when interpreting what is meant by 'sufficiently uniform'. Firstly clear genetic 'off-types' can be considered. UPOV defines these as 'plants which differ in their expression from that of the (cultivar)' [6]. This is taken by NIAB to include plants which differ in either a qualitative or quantitative character or characters. According to UPOV:

'no fixed tolerance (for numbers of 'off-types') can be determined but relative tolerance limits are used through comparison with comparable cultivars already known.' [6]

Thus the standard is set by the cultivars already in existence. The term 'comparable cultivars' is taken to mean produced by the same breeding method.

For qualitative characters, the number of 'off-types':

'should not significantly (5% probability of an error) exceed the number found in comparable (cultivars) already known.' [6]

In practice, for most qualitative characters, the number of 'off-types' in cultivars already known is very low (c.0.5%) and the tolerance limit applied by NIAB is the same as that set by UPOV for 'off-types' in self-pollinated species viz1%. Two qualitative characters, however, have higher levels of 'off-types'; these are seed:testa colour and seed:hilum colour. Although the majority of cultivars contain only an occasional testa colour 'off-type' a few have higher levels, the maximum being 15% of green seeds in two principally beige-seeded, synthetic cultivars viz. Maris Beagle and Bulldog. More cultivars, however, are mixed for hilum colour and the maximum level of 'off-types' is 35% viz. black hilums in the otherwise white-hilum, synthetic cultivar Tiger. When cultivars with these higher levels of 'off-types' are grown as controls the UPOV standard for 'off-types' in qualitative characters quoted above is applied. The question remains, however, should the new cultivar be compared with the cultivar be compared with the cultivar with the highest level of 'off-types' in the group of 'comparable cultivars' or the lowest level or some other standard? The standard decided on by NIAB as being reasonable and fair is that the number of 'off-types' should not exceed the mean number of 'off types' in the 'comparable cultivars'.

For quantitative characters including those handled qualitatively, the number of 'off-types' in cultivars already known is very low (c. 0.5%) and NIAB operates an arbitrary tolerance limit of 2%.

After any 'off-types' in both qualitative (excluding seed:testa colour and seed:hilum colour) and quantitative characters have been recorded these are excluded from further assessment of uniformity. The variability of the remaining plants, for each quantitative character, is then assessed. According to UPOV:

'the standard deviation or variance should be used as the criterion for comparison.' [6]

The new cultivars should be compared with the 'cultivars used for comparison', i.e. the controls used in the determination of distinctness. The standard operated by NIAB is that for each quantitative character the standard deviation of the new cultivar should not normally be significantly different in two consecutive growing periods, from the mean standard deviation of the cultivars used for comparison.

Stability

As far as testing stability is concerned UPOV recognises that:

'Generally, when a submitted sample has been shown to be homogeneous, the material can also be considered stable.' [6]

This view is now accepted by NIAB although formerly a second submission of seed from a different multiplication to the first was required in the second year of test and the two were grown together for comparison. Only one submission is now required and, if this is sufficiently uniform, stability is assumed to be satisfactory also.

DUS test results

The numbers of cultivars which have been tested, and the number which have failed to meet the standards for D, U or S from 1971 to 1983, are as follows:

Number of cultivars tested	Number of cultivars which have failed to meet the standards			
	NOT D	NOT U	NOT S	Total
47	4	1	2	7

DISCUSSION

Not all new cultivars meet the DUS standards. For this reason it is very important that the cytogeneticist and plant breeder must know what the standards are and take them into account in the development of new cultivars. Moreover, they need to be considered from the earliest stage in the breeding programme even from the initial choice of induced mutant plant or plant from a population, for it is in the early stages that distinctness, uniformity and stability may be decided.

In order to ensure cultivar distinctness the cytogeneticist and plant breeder should consider (i) choosing a combination of qualitative characters for which there are few or even no controls. For example, a candidate with beige testas and white hilums would have fewer controls than one with beige testas and black hilums and (ii) introducing qualitative characters which hitherto do not exist in known cultivars. Most of these can be found in the FABIS list of Genetic Variation within Vicia faba [1]. Some examples are: yellow wing spots, entirely pink flowers, yellow testas. These would be, in effect, marker genes which may serve only to improve the distinctness of the candidate. One new qualitative character which was introduced in a recently tested cultivar was the determinate growth habit: the cultivar also had diffuse wing spots (another new character). Such innovations are, as far as the DUS tester is concerned, a step in the right direction!

If, for economic or practical reasons, it is not possible or desirable to introduce new qualitative character combinations or new qualitative characters altogether the standards operated for quantitative characters should nevertheless, be borne in mind especially when breeding cultivars in an already congested group.

Attention should be given to uniformity. Uniformity problems are often eliminated by extra time being given to stabilising the cultivar before it is entered for DUS test. This ensures both uniformity and stability and in the long run may save both the breeder and the tester time and money.

REFERENCES

1. G.P. Chapman. 1981. Genetic Variation within Vicia faba. FABIS, June, pp. 12.
2. Higgins, J., Evans, J.L. and Reed, P.J. 1981. Classifications of Western European Cultivars of Vicia faba L. Journal of the National Institute of Agricultural Botany, 15, 480-487.
3. Ministry of Agriculture, Fisheries and Food. 1964. Plant Varieties and Seeds Act 1964. Part 1 - Plant Breeders Rights. London, HMSO.
4. Ministry of Agriculture, Fisheries and Food. 1973. Seeds (National List of Varieties) Regulations 1973. Statutory Instruments 1973 No. 994. London. HMSO.
5. International Bureau for Plant Taxonomy and Nomenclature (1980). International Code of Nomenclature for Cultivated Plants. Regnum Vegetabile, 104, Utrecht.
6. UPOV 1979. Revised General Introduction to the Guidelines for the Conduct of Tests for Distinctness, Homogeneity and Stability of New Varieties of Plants. Document TG/1/2. International Union for the Protection of New Varieties of Plants.

5
Interspecies Hybridisation

DEVELOPMENT OF A MODEL SYSTEM TO EXPLORE GENOME
DIVERGENCE IN THE VICIEAE

S.A. Tarawali

Department of Biological Sciences, Wye College, Ashford
Kent TN25 5AH, U.K.

ABSTRACT

Among the Vicieae, the genus Pisum is convenient since the chromosome
variation between species has been described and interspecific hybrids can
be obtained. For a given hybrid, embryogenesis will fail at a relatively
early or late stage depending on the similarity of the parents. This
paper uses these features of hybridisation in Pisum as a basis upon which
to develop, eventually, the molecular biology of both differentiation and
genome divergence.

INTRODUCTION

Differences in DNA sequence are known for chromosomes of related
species and when they occur on a sufficient scale represent a change from
homology to homeology that can be recognised by various departures from
bivalent formation at meiosis in interspecies hybrids, for example.
Larger scale chromosome differences affect earlier stages of hybrid growth
such that it may not survive even embryogenesis. Since the stages of
embryogenesis now seem likely to be associated with a precise sequence of
changing mRNA (messenger RNA) populations (Galau & Dure, 1981; Goldberg
et al., 1981; Chlan & Dure, 1983), disrupted or abortive embryogenesis in
hybrids might be reflected in a changed progression of such mRNA
populations. Theoretically, in a hybrid, mRNA progression might terminate
abruptly at a particular phase or alternatively, hybrids might from the
earliest detectable stages, show a fundamentally changed mRNA pattern
which, once recognised as being consistently associated with embryo
failure could perhaps, be used predictively. To explore this approach, a
search is being made for an appropriate model system. Within a genus,
where a graded series of relationships from close to distant can often
be recognised, conceivably, hybrid embryos could be expected to falter at
increasingly early stages.

Groups of species among which hybrids can be produced, are found
for example, in genera of the Vicieae. Interspecies hybrids have been

reported between Vicia narbonensis and its associates (Ladizinsky, 1975 and described in this book in the contributions of Ramsay & Pickersgill and of Yamamoto). Within the 40 recognised species of Cicer, C. echinospermum differs from C. reticulatum by a reciprocal translocation and a sterile hybrid can be produced between the two species (Ladizinsky, 1976). C. reticulatum and C. arietinum show cross compatability (Singh & Malhotra, 1984). A similar group of related species have been described in the genus Lens. L. culinaris is fully comptabible with L. nigricans and L. orientalis but less so with L. evoides (Cubero, 1984). For Pisum, both hybridisation data and chromosomal evidence for species differences are available (Ben Ze'ev and Zohary, 1973). Using differently related species within Pisum, a programme has been initiated permitting comparison of normal and abortive embryogenesis.

Chromosomes and interspecific hybrids in Pisum

In Pisum, four species P. sativum, P. humile, P. elatius and P. fulvum have very similar karyotypes. All have 2n = 14, P. sativum and P. humile (N) seem to have identical karyotypes; P. humile (S) and P. elatius both differ from P. sativum and P. humile (N) by a reciprocal translocation of the short arms of chromosomes IV and VI, P. fulvum is somewhat more distinct having an extra satellite on chromosome V and a much larger satellite on chromosome VII. When interspecific crosses are made these chromosome relationships are reflected. P. sativum and P. humile (N) will cross readily (reciprocally), so too do P. humile (S) and P. elatius. Crossing P. sativum or P. humile (N) with P. humile (S) or P. elatius results in hybrids with reduced gamete viability due to the presence of quadrivalents at meiosis caused by the translocation. Crosses involving P. fulvum are less successful.

A summary of these relationships is presented in Fig. 1

MALE PARENT → / FEMALE PARENT ↓	Pisum sativum	Pisum humile (N)	Pisum humile (S)	Pisum elatius	Pisum fulvum	
Pisum sativum	Fertile	Fertile	Semi-sterile	Semi-sterile	Stunted, chlorophyll deficient	P. sativum karyotype
Pisum humile (N)	Fertile	Fertile	Fertile	Fertile	Semi-sterile	Cryptic structural change
Pisum humile (S)	Semi-sterile	Fertile	Fertile	Fertile	Semi-sterile	Reciprocal translocation between short arms of chromosomes IV and VI
Pisum elatius	Fertile	Fertile	Fertile	Fertile	Semi-sterile	Cryptic structural change
Pisum fulvum	Shrunken seed	Stunted seedlings	Stunted seedlings	Stunted seedlings	Fertile	Enlarged satellite on chromosome VII, extra sacellite on chromosome V

Fig. 1   Summary of the results of interspecific hybridisations in Pisum species. Chromosome relationships are expressed for convenience as differences from P. sativum and do not necessarily imply the direction of evolutionary events. (modified from Ben Ze'ev & Zohary, 1973)

MATERIALS AND METHODS

Plant material

Seeds of Pisum humile, P. elatius and P. fulvum were obtained from the John Innes Germplasm Collection (John Innes Institute, Colney Lane, Norwich, U.K.). P. sativum seeds were the commercially available variety Hurst Greenshaft. Seeds were sown in compost in individual pots and plants raised under glasshouse conditions. (Temperature – maximum 25°C, minimum 21°C; 16 hour photoperiod.)

Embryo culture in liquid medium

Immature pods were harvested and surface sterilised for 20 minutes in 10% sodium hypochlorite, the pods were rinsed several times in sterile distilled water to reduce chlorine contamination before dissection. Subsequent operations were carried out under sterile conditions in a laminar flow cabinet. Embryos were dissected from ovules and placed in 2-

3 ml of culture medium (Stafford & Davies, 1979) in a 6-welled culture dish. Only one embryo was placed in each well. The cultures were maintained on a slowly moving orbital shaker at 20°C with a 16 h photoperiod. The medium was changed and the embryos measured every 2-3 days.

Embryo culture using nurse tissue

Embryos prepared as described above were placed on hard filter paper (Whatman no. 54) on top of apple endosperm callus growing on basic Murashige and Skoog medium without sucrose, IAA or kinetin (Flow Laboratories) solidified with 0.6% agar. The cultures were maintained at 20°C with a 16 h photoperiod. Embryos were measured every 2-3 days.

The apple endosperm callus culture was kindly supplied by Dr. David James (East Malling Research Station).

Cotyledon culture

Pods that were mature but had not started to dehydrate were harvested, the ovules were removed and placed in 20% Sodium hypochlorite for 20 minutes then rinsed well with sterile distilled water. Subsequent operations were carried out under sterile conditions. The testa and radicle were removed from each ovule and the cotyledons cut into eight pieces which were placed in 50 ml of culture medium (Davies & Cullis, 1982) in a 250 ml conical flask. The cultures were maintained on a gently moving rotary shaker at 20°C with a 16 h photoperiod.

For chromosome analysis, a piece of cotyledon was cut into fragments approximately 1 x 2 x 5 mm and hydrolysed in NHCl at 60°C for 10 minutes, the material was then squashed and stained with aceto orcein.

Polytene nuclei were isolated in a HEPES based buffer (p 17) by squashing fragments of cultured cotyledon in a petri dish. Preparation of isolated nuclei for transmission electron microscopy was as described by Chapman & Cooke (1983).

RESULTS

Embryo development

A series of distinct morphological phases can be recognised during embryogenesis. Fig. 2 shows these phases in outline, with arbitrary subdivisions, in anticipation that these will become relevant as increased understanding allows us to partition the existing major stages. Al represents fertilization and K5 is a mature, dehydrated seed. Pod dimensions vary depending on the number of ovules and are insufficiently reliable as an indicator of embryo stage. Ovule dimensions for <u>Pisum</u> <u>sativum</u> (Hurst Greenshaft) are given in the figure since these are a more useful indicator of particular stages of embryogenesis.

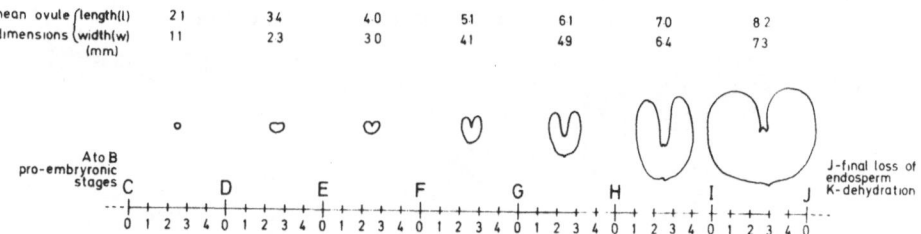

Fig. 2. Morphological change during embryogenesis in <u>Pisum</u> <u>sativum</u>.

Embryogenesis in culture

Immature embryos can be cultured in liquid medium or on solid endosperm nurse tissue. Younger embryos will survive in liquid than on the solid culture although at present, all embryos younger than stage G will lose their chlorophyll when transferred into culture. These results are summarised in Fig. 3. Modification of the culture media should allow culture of younger embryos. Embryos of <u>P. sativum</u> and <u>P. fulvum</u> can be cultured in the liquid medium.

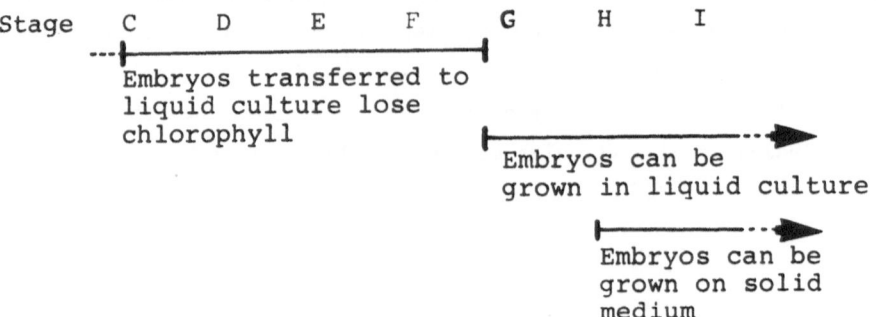

Fig. 3. Summary of the results of culturing <u>P. sativum</u>
embryos. The stages referred to are shown in Fig. 2.

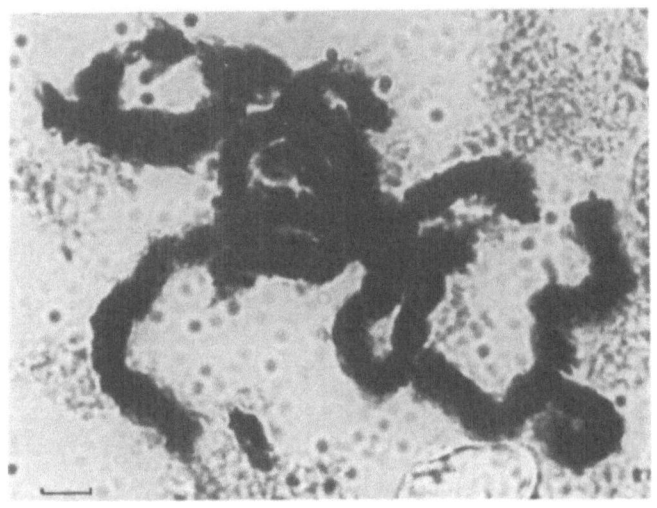

Fig. 4.   <u>P. fulvum</u> late prophase polytene nucleus from
cotyledon culture. Optical microscopy, aceto-orcein stain.
Bar = 5.0 μm

Late embryogenesis - cotyledon culture

Towards the end of embryogenesis the cotyledons are swollen and
full of storage protein. Cell division has ceased at this stage but
nuclear replication still occurs giving rise to polytene nuclei. The
proportion of polytene nuclei can be substantially increased by placing
fragments of cotyledon in a liquid culture containing 2,4-D. After 4-7
days the majority of cotyledon nuclei are seen to be polytene and in
prophase (Fig. 4). Irregular chromatin strands can be seen and often it
is possible to identify connections of satellite chromosomes with the

nucleolus organiser region. Using techniques developed in this laboratory by P.M. Allington, the polytene nuclei can be isolated and viewed using the transmission electron microscope (see p 25). Recently it has been possible to culture embryos in liquid medium to a sufficiently mature stage such that when transferred to a "polytene culture medium" (as described) the nuclei of the cotyledon cells become polytene (Fig. 5).

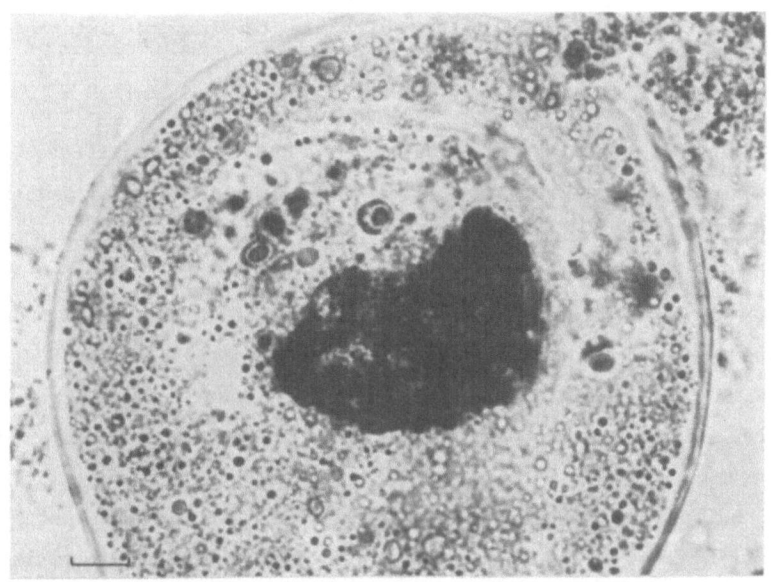

Fig. 5. Aceto-orcein stained squash preparation of a cotyledon cell from P. sativum showing a polytene nucleus at prophase. The cotyledon material was taken from an embryo grown in liquid culture from stage G onwards (see Fig. 2), then cultured in "cotyledon culture medium" containing 2,4-D for 4 days. The nucleus is c. 3 times the diameter of nuclei prior to culture in 2,4-D. Bar = 10 µm

DISCUSSION

The later a normal embryo is removed from the seed, the greater its metabolic autonomy. The more its dependence in the earlier stages can be accommodated *in vitro* the better understood is that dependence. If exogenously, the medium were complete and the environment at optimum, presumably under such conditions, clusters of appropriate mRNAs would appear on cue throughout an orderly embryogenesis. This arrangement, however, even at best, must be a fragile one that designedly, in our approach, might be disrupted by endogenous causes. A hybrid embryo with functionally incompatible parental genomes would eventually abort. This paper proposes the concept that a collection of 'experimental zygotes' each known to abort at an individual and precise point in embryogenesis offers a stepwise approach to the biochemistry of development. This if successful would both characterise normal events and identify the point of breakdown and possibly, a cause of genomic incompatibility for a hybrid.

Results obtained so far, together with what was already known about Pisum suggest that it is potentially a useful model system with wide application but with particular relevance to Vicieae for the following reasons.

1. A range of readily available species has been identified. These can be crossed to yield hybrids of varying viability, depending on the species involved. Moreover, chromosomal differences among these species have been described in some detail.

2. Embryos can be dissected from immature ovules and cultured under sterile conditions. This provides an opportunity to study in vitro normal and abortive embryos and if necessary to be able to remove them at precisely defined stages of development easily and without damage.

3. As culture techniques improve, there is the prospect of 'rescuing' hybrid embryos that might abort in vivo and resolving the possible causes of abortion.

4. Pisum polytene chromosomes can be readily induced using cotyledon culture. Assuming a similar situation to that with dipteran salivary gland chromosomes then there would be substantial lateral replication of gene sites including presumably, those of what in a unineme, would be single copy genes. (From work in this laboratory it is already known that polyteny can be induced in Vicia faba.)

5. There now exist for Pisum, 'gene libraries' of nucleic acid probes (see Lycett et al., 1983) and some of these would be sufficiently 'broad spectrum' for use not only with Pisum but other genera in Vicieae.

6. Where hybrid embryos can develop to the late cotyledon stage, there is the prospect of nuclei acid probing of polytene chromosomes in such cases.

One further point is worth consideration. Allowing for the differences in multiples of DNA content known for individual nuclei in different species of Vicia for example, a molecular approach makes it possible to extend comparison beyond the limits of sexual accessibility. Given a sufficient input of resources it should eventually be possible to establish whether the presently insurmountable barriers to hybridity within Vicieae isolate what are largely similar genomes.

Although here, emphasis has been placed initially on Pisum, a key factor in this approach is that of a study the methodology of which is applicable throughout the Vicieae.

REFERENCES

Ben Ze'ev, N. & Zohary, D. 1973. Species relationships in the genus Pisum L. Israel J. Bot. 22, 73–91.

Chapman, G.P. and Cooke, S.A. 1983. A technique for optical and electron microscopy of plant chromosomes. Protoplasma 116, 198–200

Chlan, C.A. & Dure, L. 1983. Plant seed embryogenesis as a tool for molecular biology. Molecular & Cellular Biochem. 55, 5–15.

Cubero, J-I. 1984. Taxonomy, distribution and evolution of the lentil and its wild relatives. In Genetic Resources and Their Exploitation – Chickpeas, Faba beans and Lentils. eds. J.R. Witcombe and W. Erskine. Martinus Nijhoff/Dr. D.W. Junk Publishers, The Hague. pp. 187–203.

Davies, D.R. & Cullis, C.A. 1982. A simple plant polytene chromosome system and its use for in situ hybridization. Plant Mol. Biol. 1, 301–304.

Galau, G. & Dure, L. 1981. Developmental biochemistry of cottonseed embryogenesis and germination: changing messenger ribonucleic acid populations as shown by reciprocal heterologous deoxyribonucleic acid – messenger ribonucleic acid hybridization. Biochem. 20, 4619–4178.

Goldberg, R.B., Hoschek, G., Tam, S.H., Ditta, G.S. & Breidenbach, R.W. 1981. Abundance, diversity and regulation of mRNA sequence sets in soybean embryogenesis. Dev. Biol. 183, 201–217.

Ladizinsky, G. 1975. On the origin of the broad bean Vicia faba L. Israel J. Bot. 24, 80–88.

Ladizinsky, G. 1976. Genetic relationships among the annual species of Cicer L. Theor. & Appl. Genet. 48, 197–203.

Lycett, G.W., Delauney, A.J., Gatehouse, J.A., Gilroy, J., Croy, R.R.D. and Boulte, D. 1983. The vicilin gene family of pea (Pisum sativum L.): a complete cDNA sequence for preprovicilin. Nuc. Acid Res. 11, 2367–2380.

Singh, K.B. & Malhotra, R.S. 1984. Collection and evaluation of chickpea genetic resources. In Genetic Resources and Their Exploitation – Chickpeas, Faba beans and Lentils, eds. J.R. Witcombe and W. Erskine. Martinus Nijhoff/Dr. W Junk, Publishers, The Hague. pp. 105–122.

Stafford, A. & Davies, D.R. 1979. The culture of immature pea embryos. Ann. Bot. 44, 315–21.

INTERSPECIFIC HYBRIDISATION IN VICIA SECTION FABA: COMPARISON OF
SELFED AND HYBRID EMBRYO AND ENDOSPERM

Gavin Ramsay and Barbara Pickersgill

Department of Agricultural Botany, University of Reading

INTRODUCTION

In certain interspecific combinations involving Vicia faba (e.g. V. faba var. paucijuga x V. johannis var. procumbens) fertilisation occurs frequently but embryos do not reach a sufficient size to be cultured by conventional techniques. To improve chances of rescuing and successfully culturing these very small embryos it is necessary to determine how far, and for how long, they will develop in vivo. In addition, comparison of rates of growth of embryo and endosperm in inter- and intraspecific crosses may lead to an understanding of the reasons for hybrid embryo abortion. Ways of delaying embryo abortion can then be explored.

METHODS

Inter- and intraspecific crosses were made and pistils fixed 1, 2, 3, 5, 7, 10, 14 and 21 days after pollination. Ovaries were embedded in paraffin wax, sectioned and stained using the Feulgen method. Numbers of nuclei (n) in endosperm and embryo were counted and the numbers of division cycles were calculated ($\log_2 n$ for endosperm; $[\log_2 n] + 1$ for embryo).

RESULTS

Intraspecific V. faba embryos (Fig. 1) grew at a logarithmic rate for about 21 days, by which time they contained about 250,000 cells. The mean cell doubling time (MCDT) over this period was 21.6 hours. Root-tip MCDTs lie between 13.0 and 18.1 hours (Bennett et al. 1972; Caryologia 25, 445-453). Endosperm MCDT was similar to that of embryos for the first 7 days but soon declined. At 21 days, the endosperm was coenocytic and contained about 5,000 nuclei.

Interspecific crosses between V. faba and V. johannis (fig. 2) gave embryos which grew more slowly than intraspecific ones. Growth did not maintain a logarithmic rate but slowed after 5 days. Micronuclei, which may indicate chromosome elimination, were rarely found. The largest embryos, which consisted of about 50 cells, were found from days 10 to 14. No embryos survived as long as 21 days.

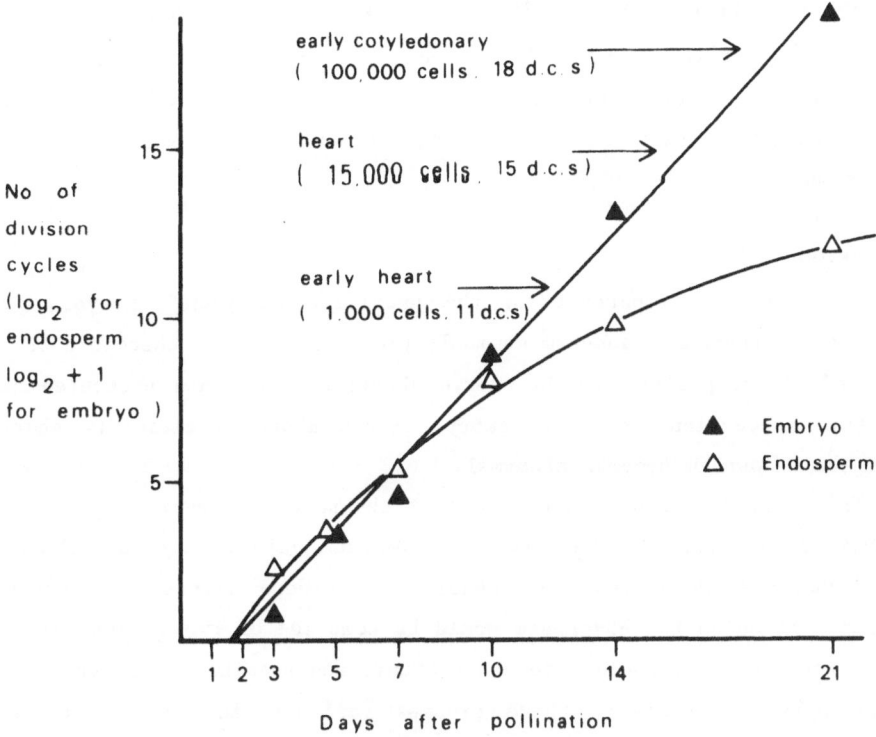

Figure 1. Embryo and endosperm growth rates: Crosses
within <u>V. faba</u> var. <u>paucijuga</u>.

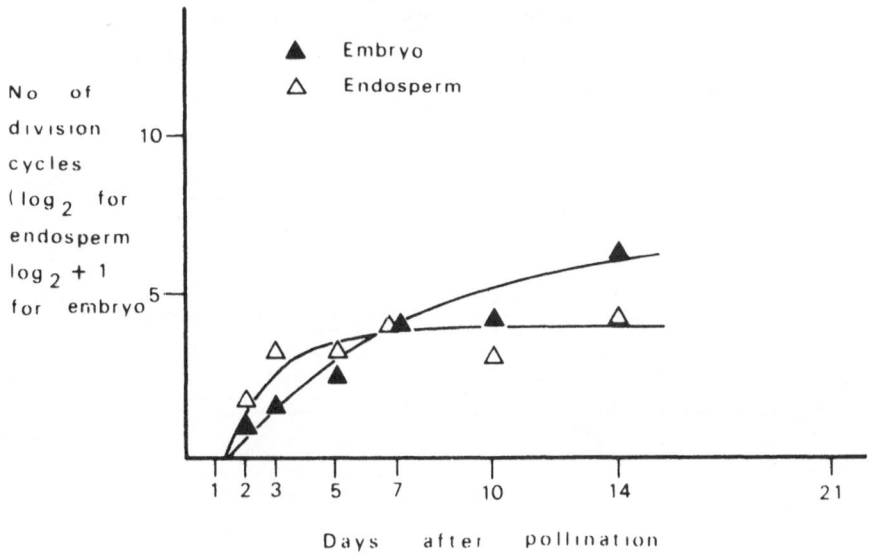

Figure 2. Embryo and endosperm growth rates: <u>V. faba</u>
var. <u>paucijuga</u> x <u>V. johannis</u>.

Hybrid endosperm nuclei divided quickly at first but the maximum number of nuclei (about 16 on average) was reached at day 5. Subsequently, large, irregular nuclei were produced through endomitosis. These contrasted with the small, regular nuclei of the intraspecific endosperm.

DISCUSSION

Endosperm is important in the nutrition of young embryos. The presence of abnormal endosperm commonly precedes, and may therefore be one cause of, interspecific hybrid embryo abortion. The data presented here fit this suggestion as hybrid embryo growth slows dramatically shortly after the endosperm becomes abnormal.

The very different DNA contents of the parental species (V. faba 27 pg DNA/2C nucleus; V. johannis 14 pg DNA/2C nucleus) may contribute to the disharmony which leads to abnormal endosperm mitoses. Reciprocal crosses, in which the endosperm would be composed of two V. johannis and one V. faba genomes, would have a different, and possibly more harmonious, genome balance. However, these crosses fail due to poor pollen tube growth. Induced tetraploids of the wild species might cross with diploid V. faba to produce more balanced hybrid nuclei. Alternatively, the use of species which are not included in section Faba but which have DNA contents more similar to that of V. faba may produce hybrid endosperm capable of completing more division cycles. V. melanops (20 pg DNA/2C nucleus) is the best candidate in this context and has already been shown by us to be capable of fertilising ovules of V. faba.

A NOTE ON INTERSPECIFIC HYBRIDIZATION BETWEEN VICIA NARBONENSIS AND ITS RELATED SPECIES

Kiyoshi Yamamoto

Faculty of Agriculture, Kagawa University,
Kagawa-Ken, Japan.

In order to clarify the genetic relationship between species in the section Faba of the genus Vicia, karyotypic and isoenzymatic studies were carried out. The species examined were V. narbonensis, V. serratifolia, V. galilaea, V. johannis, and V. hyaeniscyamus known as the V. narbonensis species group, together with the species V. faba and V. bithynica.

Considerable differences were observed between the karyotypes of the species in the group of V. narbonensis and those of the species V. faba and V. bithynica, in regard to the relative length of satellite chromosomes and the ratio of short arm to long arm of each chromosome. Among the species of the V. narbonensis group, some variations were observed in the size of the satellite chromosome and the shape of submedian chromosomes.

Isoenzyme patterns and morphological characters also varied among the species. However, based on the observed karyotypes, morphological characters and the isoenzyme band patterns, clearer genetical relationships were deduced for the species examined. It is presumed that V. faba, V. bithynica and the V. narbonensis species group, are remote from each other. It is suggested, however, that V. galilaea is the most evolved in the V. narbonensis group, and is an intermediate between this group and between V. faba. Within the V. narbonensis species group, a more closer relationship is suggested between the pairs: V. narbonensis - V. serratifolia and V. johannis - V. hyaeniscyamus.

Interspecific crosses between Vicia narbonensis ssp. and V. faba. V. narbonensis ssp. and V. bithynica were carried out. Despite many variations of technique, no hybrids were obtained. Among V. narbonensis and its related species F_1 hybrids and their progeny were obtained. Cytogenetical studies were carried out on these hybrids.

In F_1 plants feeble growth was observed except the F_1 of V. johannis x hyanecyamus because F_1 plants were chlorotic. Fertility of F_1 plants was low except the F_1 in the case mentioned. In the F_1 plants of this hybrid type normal growth was demonstrated and the fertility was high. Seven normal bivalents were observed in first meiotic metaphase of these plants. In the other combinations, many types of chromosome figurations with univalents and multivalents were observed. Additionally, much variation was observed among the different lines of V. narbonensis as regards the karyotype and isoenzyme both for band patterns and morphological characters. In 23 lines of this species, three karyotypes were demonstrated as A, B and C. The difference was observed in chromosomes I, II, V and VI in the length, the ratio of short arm to long arm length, but not so much difference was observed in the SAT chromosome and the total length of chromosome set.

Large differences were observed in the isoenzymic band patterns among the lines of this species. In this regard strains of karyotype B and C were not different from type A. There appeared to be little relation between the karyotype and the corresponding morphological characters.

It was concluded that V. faba is distant genetically from V. narbonensis ssp., as well as distant from V. bithynica in the karyotypic, isoenzymic and morphological characteristics studied. Interspecific hybridization results between the species of the section Faba tend to support this.

Closer genetical relationship was presumed in V. narbonensis ssp. Of these, the closest were sub-species johannis and hyaenicyamus.

Evolutionary trends are not clear, but the furthest genetical distance appeared to be between V. faba and V. bithynica.

6
Haploidy

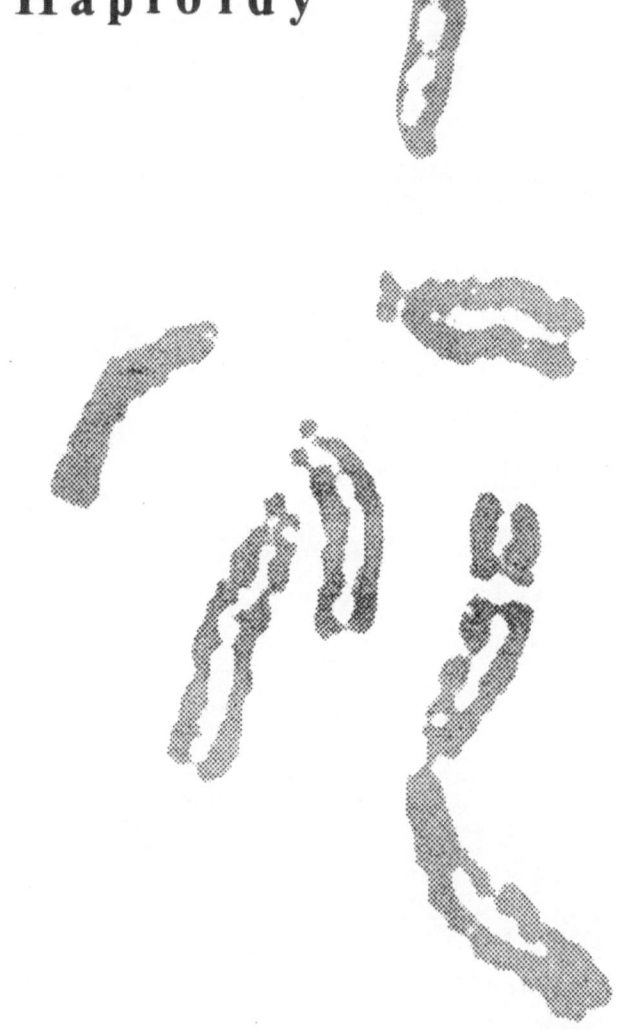

A NOTE ON HAPLOIDY

T. Paratasilpin

Chiang Mai University,
Chiang Mai, Thailand.

The most effective medium to sustain pollen growth and development (Fig. 1) was Murashige and Skoog (1962) inorganic salt solution with a modified vitamin amino acid solution (meso-inositol 100mg/l; glycine 2mg/l; thiamine HCl 0.5 mg/l; nicotinic acid 5.0 mg/l; pyridoxine HCl 0.5 mg/l; folic aid 0.5 mg/l; biotin 0.05 mg/l). Multinucleate pollen grains were initiated in the medium containing kinetin (0.2 mg/l) although the absolute concentration was not found. The production increased considerably when either NAA (0.1 mg/l) or pCPA (0.1 mg/l) or 2,4-D (5.0 mg/l) was also added to the medium whereas when pCPA or NAA were used alone, only a few grains divided. 2,4-D apparently inhibited pollen division, either supplementing alone or in combination with other auxins in the medium, but enhanced the effect of kinetin. Initially, 2,4-D was also considered to involve with internal wall formation of a multinucleate grain but later results did not confirm this. When NAA (0.1 mg/l) was used in a medium, a higher level of kinetin could be used while a lower level was more suitable when pCPA was supplemented in the medium for multinucleate grain formation. Three per cent sucrose appeared to be the most effective concentration for pollen growth.

The rapid increase in the percentage of empty grains up to 100 per cent within five to six days after culturing could be interpreted as the interaction between the culture medium and the original physiological status of the pollen grains and not the result of plasmolysis due to the high concentration of the medium as a whole. Whether or not zeatin and benzylaminopurine could replace kinetin for pollen growth is uncertain.

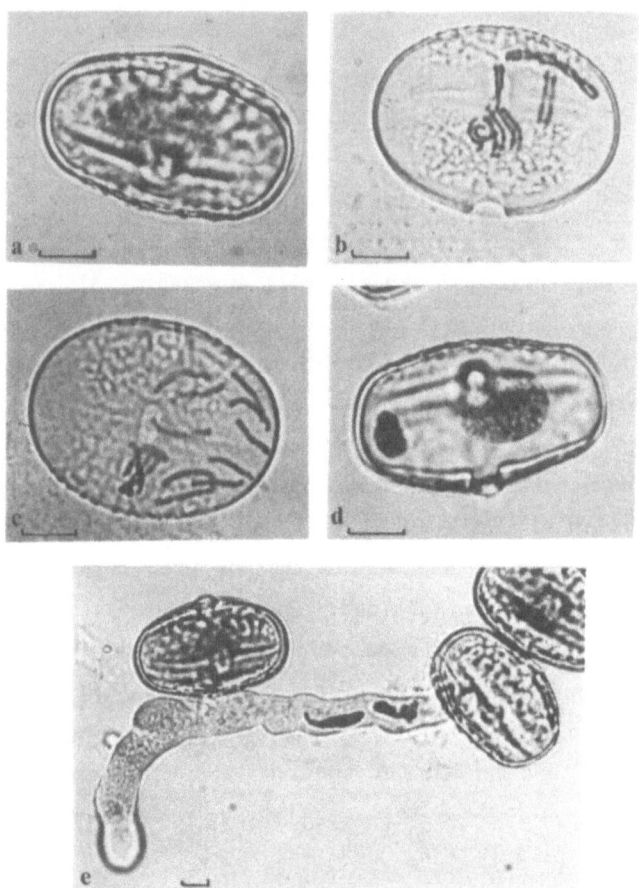

Figure 1. Developmental sequence in a V. faba pollen
grain following its release from a tetrad as a uninucleate
microspore (a); (b) the microspore at metaphase and (c)
anaphase. Eventually binucleate pollen is formed (d).
After germination, two male gametes resulting from
division of the generative nucleus move into the pollen
tube (e). Bar = 10 μm.

Compacta appeared to be the most promising cultivar of V. faba to
produce pollen capable of dividing repeatedly. The uninucleate stage of
pollen was observed to be the most plastic stage which, under a suitable
condition, could be switched from its normal role into the repeated
division of the nucleus. Multinucleate pollen contained mostly, the
vegetative derivatives up to eight nuclei while the division of a

generative nucleus was limited to two or three (Fig. 2). A proembryoid containing seven cells (Fig. 3) was observed when the medium contained kinetin (0.2 mg/l), NAA (0.1 mg/l) and 2,4-D (5.0 mg/l), but no further development was found. Cold pretreatment to flower buds which was found to increase the number of grains entering the first mitosis and the number of symmetrical binucleate grains did not show a consistent effect on stimulating multinucleate grain formation.

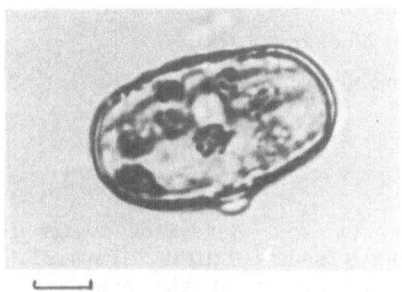

Figure 2.   8 nucleate proembryoid.   Bar = 10 μm.

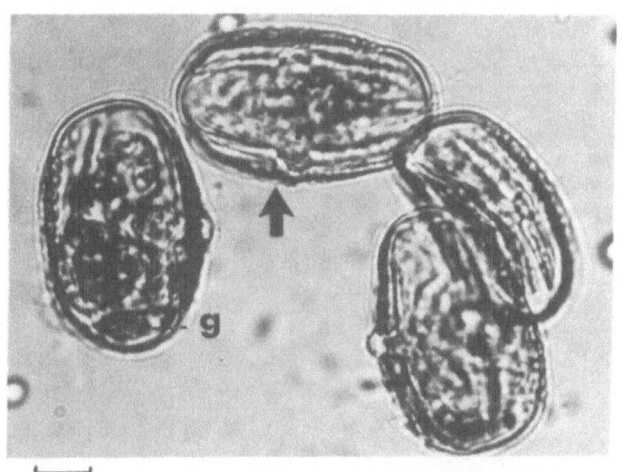

Figure 3.   A proembryoid containing 7 cells, one of which
was the generative cell (g). A microspore (large arrow)
still remained quiescent (n = nucleus).   Bar = 10 μm.

The original physiological status of pollen before culturing was considered to be a main factor influencing pollen development in culture. Abnormalities of pollen formation were observed when the environmental temperature was high but the other abnormalities in microsporogenesis, resulting perhaps in the upset of a normal physiological status of pollen which visually undetected (Rowlands, 1958), could also take place when only a small variation of environmental conditions arose. Both factors were believed to influence also responses of pollen in culture, particularly proembryonic grain production. However, it was considered that anthers harvested in May at Wye (i.e. early summer) could give the most promising result.

ABBREVIATIONS

NAA   Napthalene acetic acid

pCPA   p-chlorophenoxyacetate

2,4-D 2,4-dichlorophenoxyacetic acid

REFERENCES

Murashige, T. & Skoog, F. (1962). A revised medium for rapid growth and bioassays with tobacco tissue cultures. Physiol. Plant. 15, 473-497.
Rowlands, D.G. (1958). The nature of the breeding system in the field bean (Vicia faba) and its relationship to breeding for yield. Heredity 12, 113-125.

7
Mutant
Physiology

A NOTE ON TWO CONTRASTED GROWTH FORMS OF FABA BEAN

M.P. Bailey

Department of Biological Sciences, Wye College, Ashford,

Kent. TN25 5AH, U.K.

In the last decade a new type of field bean, a 'cereal mimic' with a determinate growth habit, has been introduced into the breeding pool. It is derived from an x-ray mutant from the Swedish line Primus with a terminal inflorescence, ti, which breeds true as a homozygous recessive (Sjodin, 1971). This has been crossed with a precocious flowering genotype. The resulting plant is described as neotonous because it flowers while it still has only juvenile leaves (Chapman and Peat, 1978). The combination approximates to a proposed ideotype recalling the cereal growth habit.

Maris Bead grows to a maximum height of 1700 mm with typically three main stems whereas the determinate line studied grows to a maximum height of 300 mm and produces numerous side branches (there is no significant difference between the types in final total stem length). The determinate type also has a lower leaf area index and leaf area duration. The leaves last longer before abscission than those on the indeterminate plant, and are subjectively darker, thicker and tougher. Some of these differences are due to the neotonous leaf form, induced by the combination of determinacy and precocious flowering, and some directly to the determinacy.

This project is a study of the differences in assimilate production and partitioning between the determinate and indeterminate forms, which, because of the relative newness of the determinate, have been little studied to date (Austin et al., 1981; Baker et al., 1983 and in press).

All date refer to glasshouse grown plants at low planting density.

The initial approach to the problem was to characterise the two plant types by measuring leaf area, weight, chlorophyll content and photosynthetic capacity of leaves of different ages and at different positions on the main stems of the plants. Maris Bead was selected as a 'typical' indeterminate variety and the line 19280/1 as a 'typical' determinate. (In fact 19280/1 is a rather extreme determinate plant, many lines are significantly taller.)

(Chlorophyll content was measured by extraction in acetone, partitioning into petroleum spirit and spectrophotometry at 645 and 663 nm. Photosynthetic capacity was measured by feeding leaf discs with $^{14}CO_2$ [30 min in 0.08% CO_2 0.52MBq/1], solubilising and liquid scintillation counting.)

From these, derived attributes such as specific leaf area were calculated.

| | 1982 | | 1983 | |
	DETERMINATE	INDETERMINATE	DETERMINATE	INDETERMINATE
Attribute Durations				
Area (m^2). days	7.6	28.4	1.8	9.8
Dry Weight (g.days)	207	946	56	290
Chlorophyll Content (mg.days)	3.4	14.9		
Photo-activity (kBq.days)	8.6	23.1		
Seed Numbers				
Seeds	14.02 (2.50)	47.00 (9.28)	15.71 (1.30)	38.92 (1.98)
Seed Dry Weight (mg)	382 (8)	473 (9)	260 (5)	358 (7)
Total Seed DW on Plant (g)	5360	22200	3880	15000
Yield Efficiency (Seed weight per ...)				
per m^2 day	706	782	2156	1535
per g DW.day	25.9	23.5	69.3	51.9
per mg Chlorophyll.day	1560	1490		
per kBq Photo-activity.day	624	962		

(Standard errors in brackets).

Table 1. Yield efficiency calculations (adapted from Baker et al., in press).

For each attribute/stem combination a regression surface was fitted of attribute value against leaf age and node number. By adding other data such as mean plant age at leaf opening and mean leaf age at abscission etc., these surfaces were used to reconstruct a model plant. The models were checked back against less detailed whole plant measurements.

From these model plants leaf area durations etc. were calculated. These are shown in the top section of Table 1.

Since the main purpose of growing field beans in the UK is the production of seed these durations have been compared with seed yield and an assessment of seed production efficiency made as shown in Table 1.

In 1982 the plants were well watered but in 1983 water stress was very apparent; leaf area duration of both types was cut by 75%. However, seed yield was cut by only 30%. In 1982 yield efficiencies of the two forms were very similar whereas in 1983 the efficiency of the determinate plant was some 40% greater than that of the indeterminate.

The next step of the project was to study photosynthetic activity of attached leaves (using an IRGA) and to measure dry weight partitioning.

It was found that the photosynthetic activity of determinate leaves was 24% higher than that of indeterminate ones (when extrapolated back to a leaf age of 0 to eliminate the effects of the longer lifespan of determinate leaves). These measurements were conducted in 1983 and compare well with the yield efficiency calculations.

Dry weight distributions in the two forms were similar except that the determinate pattern was compressed in time because of the shorter life cycle. These results are comparable with the harvest indices reported by Baker et al. (1983) which are measured on similar material in earlier years.

In 1984 experiments are being continued with measurements of dry weight distributions in plants grown at various planting densities (4 to 100 plants/m^2 for indeterminate, 4 to 150 plants/m^2 for determinate). A minor part of this will be an investigation of the micro-climate within the canopy; an interesting area for pathologists.

Conclusions from the work so far are that the leaves of determinate and indeterminate plants are more or less equally efficient at producing assimilates for seed growth under good growing conditions but that the determinate form is possibly less prone to variability in yield under less ideal conditions (as shown by the smaller reaction to water stress in 1983). When added to the advantages of a low plant with a short life cycle (such as reduced lodging, aphid avoidance, ease of harvesting, better compatability with a cereal based agronomy, ease of pesticide spraying etc.) this should bode well for the future.

However, the particular determinate variety studied here is quite simply too small, with a leaf area duration of only a quarter that of the indeterminate crop. It also suffers from infertile (or late fruiting) side branches, in much the same way as the indeterminate form has an infertile stem apex. Whether side branch production will be efficiency suppressed by high density planting is not yet known but it may be equally easy to breed it out of the crop.

It is known that many features of the determinate growth habit are dependent upon the genetic background upon which it is superimposed. Similar studies to that reported here on a range of determinate lines would now appear worthwhile.

From all these considerations it would appear that a promising target for breeding would be a determinate form with perhaps 15 leaf nodes on a single stem, the last four of which should flower simultaneously.

REFERENCES

Austin, R.B., Morgan, C.L. and Ford, M.A. 1981. A field study of the carbon economy of normal and 'topless' field beans (Vicia faba). In : World Crops: Production, Utilization, Description. Vol. 4 Vicia faba: Physiology and Breeding, (Ed. Thompson, R.) p. 60, Martinus Nijhoff Publishers, The Hague.

Baker, D.A., Chapman, G.P., Standish, M. and Bailey, M.P. 1983. Assimilate partitioning in a determinate variety of field bean. In : Temperate Legumes: Physiology, Genetics and Nodulation (Ed. Jones, D.G. and Davies, D.R.). Pitman Books Ltd., London.

Baker, D.A., Chapman, G.P., Standish, M. and Bailey, M.P. (in press) Partitioning of assimilates in the determinate and indeterminate forms of faba bean. In : Proceedings of Vicia faba General Review Symposium, Sutton Bonington, Sept. 1983.

Chapman, G.P. and Peat, W.E. 1978. Procurement of yield in field and broad beans. Outlook on Agriculture 9, 267-272.

Sjodin, J. 1971. Induced morphological variation in Vicia faba. Hereditas 67, 155-180.

PROGRESS AND PROBLEMS IN BREEDING FOR HIGH SEED PROTEIN
THROUGH MUTATIONS IN VICIA FABA IN EGYPT

A.M.T. Abo-Hegazi

Radiobiology Department, Nuclear Research Center,
Atomic Energy Establishment, Cairo, Egypt.

ABSTRACT

The important problems facing programs for improving quality and quantity of seed proteins (as well as other characters) of faba bean include lack of a wide base of natural variability of Vicia faba L., negative correlation between seed protein content and seed yield, susceptibility of V. faba seeds containing high protein to bruchids such as Callosobruchus maculatus F., sharp deterioration of seed yield by selfing, high percentage of natural crossing beside the contradiction noticed in results of previous publications on variability and correlations of seed protein content and seed yield. However, mutants showing high seed protein content with nearly equal or a little less seed yield have been produced through the use of doses ranging from 0.5 to 10 kr of CO^{60} gamma rays. Results of successive generations (cycles) of selection for high seed protein are given.

INTRODUCTION

Faba bean (Vicia faba L.) is one of the oldest pulse crops grown in Egypt. It was thought to be grown in ancient Egypt but for some unexplained reasons it was regarded as an unworthy food by the ancient Egyptians and hence does not appear in tombs. The priests would not eat it and left it disdainfully to the common people (Aykroyd and Doughty, 1964).

Nowadays it occupies over 120,000 hectares producing about 252,000 tons of seeds (Khalil, 1984), an amount about equal to local consumption.

Some contradiction was found between previous results concerning variability existing in local field bean having regard to the fact that most of the experiments were carried out on the same local material. Ibrahim (1954) found that selection was not recommended with local ecotypes of field bean due to lack of variability. This was confirmed by the same investigator in 1963 and 1981 where he reported on progress in selection programmes that have been started as early as 1929. Abdalla (1981) mentioned that "breeders know that this crop suffers from a narrow useful variability. This is why different problems associated with the production of this crop have remained unsolved."

On the other hand, Abdalla (1964) found that "variability was very striking and offers a wide base for crop breeders to select better varieties from the local stocks." This was confirmed by Shalaby (1965); Ali (1970) who also reported no correlations between seed yield and other characters and Mitkees and Hassan (1983) who found good possibilities for successful selection for high protein content of the seed. Hawtin (1981) from outside Egypt, reported on the wide variability which exists in field beans. Nevertheless, to the knowledge of the present author, no one improved variety has been produced (selected) from the local material neither in the Ministry of Agriculture nor in other institutes which put some weight to results and conclusions of the first group of investigators.

Seed protein is a very important character from the point of view of nutrition as well as for taxonomic reasons, i.e "populations could be best characterized and discriminated by pattern of seed protein and esterases", as studied by Kaser and Steiner (1983).

Efforts for increasing seed protein content of this crop in Egypt have followed different ways such as agronomic treatments in the form of nitrogen application (Rabeia, 1977), increasing rate of phosphorus application (P_2O_5) (Metwally, 1973 and Rabeia, 1977), and increasing the number of irrigations (Metwally, 1973), inoculation with Rhizobium (Saber, 1980; Tolba, 1980 and Abo-Hegazi et al., 1981), spraying seedlings or plants that developed from irradiated or non irradiated seeds with certain concentrations of various growth regulators and some other chemicals such as salycilic acid and indole acetic acid (Ateia, 1981), nicotinic acid and nicotinamide (Rabeia, 1977), Trifluralin (triflan), nitralin, 2,4-DB (2,4-di-chlorophenoxy buteric acid), Fluometuron (Cotoran), P = amino benzoic acid, Folifertil compound (FF), (El-Batal, 1979), introduction of new germ-plasm and hybridization with local varieties as proposed by Ibrahim (1954, 1963), or the induction of mutations by means of various mutagens such as gamma rays (Abo-Hegazi, "since" 1973; Hassan, 1977; El-Kady, 1978), EMS alone or combined with gamma rays (Hussein and Abdalla, 1978, 1979).

Studies for understanding the mode of inheritance of seed-protein showed the absence of dominance. The estimated number of genes governing this character were four or five gene pairs, two of which produced the major effect and their action was more additive. Heritability estimates ranged from 45.74 to 80.00% showing that individual plant selection for

high protein content could be practiced successfully, Omar et al. (1979).
However, dominant gene action was prevalent for protein as shown by
Mitkees and Hassan (1983) where they found that environmental variance was
21% and additive variance was 24%, suggesting a possibility for successful
selection for this character. Hassan (1982) found that protein content
showed equal importance of both additive and dominance types of gene
action, as well as epistatic effects. Abo-Hegazi (1979a) found that
heritability estimates of this character were ranged from 32.23 to 70.33%
showing the existence of a high portion of genetic variance for this
character. Weather conditions (maximum, minimum, difference between them
and average temperatures) may affect this character as shown by Abo-Hegazi
et al. (1978).

Unlike the situation usually noticed in grain, studies on
correlations between seed protein content and seed yield in field bean
revealed the absence of significant correlation according to Bond (1977),
Griffiths and Lawes (1978), de Vries (1979) cited after Sjodin (1981).
The same trend was also noticed by Abdalla et al. (1976), Abdalla (1979)
and Hussein and Abdalla (1979) on Egyptian stocks. Picard et al. (1981)
expected high yielding F_1 varieties with high protein content. On the
other hand, negative and strong correlation was found between seed yield
and seed protein content (Abo-Hegazi, 1975). This was confirmed later on
a large number of Egyptian and imported varieties and lines of field bean
(148) when tested in northern, middle and southern Egypt (Abo-Hegazi et
al., 1978). The same situation has been reported in India (Anonymous,
1971).

Correlation between high content of protein in seeds of field bean and
high susceptibility of seeds with some bruchids such as Callosobruchus
maculatus F. Natural and artificial infestation was noticed by Abo-Hegazi
and Ahmed 1982).

Results of El-Sherbeeny (1970) proved that field bean plants stand
mid-way between the two extremes of self- and cross-fertilized crops. The
average percentage of natural crossing was 63.32%. Distance between
parents and distance to apiaries (honey bees) strongly affect this
average. Seasons had slight influence upon the amount of out crossing.
Selfing of field bean plants leads to strong reduction in the yield of
seed. On the average, the open pollinated plants yielded nine times
compared with bagged plants.

The above mentioned may represent the main constraints on programs aiming to improve characters like seed protein in field bean.

Improving the quantity and quality of seed protein of this crop will need a tool for increasing variability of local varieties. Therefore, doses of CO^{60} gamma rays from 0.5 to 10 krad have been utilized by Abo-Hegazi (1973a) to increase the variability of this character as well as other plant characters of local varieties of field bean. Hassan (1977), El-Kady (1978) and Hussein and Abdalla (1978 and 1979) utilized either gamma rays and/or some chemical mutagens for the same purpose.

MATERIALS AND METHODS

The material used in the present studies are <u>V. faba</u> seeds born on plants originated in the fourth generation after seed irradiation of various doses of gamma rays (Cycle I) and open pollination in the next generation (Cycle II).

Screening for seed protein content was carried out on a large population of plants resulted from <u>V. faba</u> seeds previously irradiated with 0.5, 1.5, 2, 2.5, 3, 3.5, 4, 4.5, 5 and 10 kr of CO^{80} gamma rays. DBC (Udy, 1971) methods were used for the screening. Selection was carried out for individual plants which possess 38% or more of DBC-crude protein in their seeds.

RESULTS

Results on one of the experiments carried out within the framework of our mutation breeding programme for increasing seed protein content of <u>Vicia faba</u> will be included here.

Results on two successive cycles (generations) of selection for high protein in the seed, are presented in Table 1. It is shown that the frequency of plants containing high protein was 0.8% in Cycle I in the population resulting from unirradiated seeds. This frequency was increased to reach 1.5% in progenies of 3.5 kr treatment. Therefore, some of the used doses increased the frequencies of plants showing high content of crude protein in their seeds. Respective frequencies were increased in Cycle II, that is increased frequencies due to both of selection and gamma ray treatments. In this connection, it should be mentioned that Cycle II was done under open pollination because tripping did not increase yield in the present studies.

Dose	Percentages of plants	
(Krad)	Cycle I	Cycle II
Control	0.8	0.9
0.5	0.7	0.8
1.0	1.1	1.1
1.5	0.6	0.6
2.0	0.9	1.3
2.5	0.5	0.9
3.0	0.7	1.0
3.5	1.5	1.5
4.0	0.8	1.4
4.5	1.1	1.1
5.0	1.5	1.3
10.0	0.8	0.7

Table 1. Frequencies of plants containing 38% or more of crude
protein in their seeds in populations of faba bean (Vicia
faba L.) resulted in two generations (cycles) after seed
irradiation with doses of CO-60 gamma rays.

Mutants and mother variety	Dose (krad)	Seed - protein content	Seed yield
Mother Var. G2	0.0	100.0	100.0
M 378	5.0	112.4	97.2
M 1011	4.0	109.6	105.2
M 257	0.5	134.9	96.3
M 556	1.0	116.2	100.4

Table 2. Seed yield, seed protein content and original gamma
ray dose of mutants with high content of seed protein.

The programme yielded a small number of mutants showing increases in
crude protein content which reached 34.9% of the original variety with
seed yield 3.7% less than the parent variety as shown in Table 2.

DISCUSSION

It may be of value to mention that seed yield in the majority of selected high protein plants was not sufficient to be transferred to the next generation due to negative strong correlation observed between seed yield and seed protein content, (Abo-Hegazi, 1975; Abo-Hegazi et al., 1978 and Anonymous, 1971). Selfing of the selected high protein plants was an additional major factor for no seed setting at all or a very few seed yield (El-Sherbeeny, 1970). Selfing is essential in breeding programmes for maintaining lines possessing any character such as high content of seed protein. However, some of the selected plants were kept under open pollination in the field. But, seeds and pods resulted on selected plants (selfed and non selfed) were usually infested with some bruchids causing additional losses in selected "valuable" seeds (Abo-Hegazi and Ahmed, 1978 and Abo-Hegazi, 1982). Therefore, correlations between seed protein on one hand and seed yield (negative) and susceptibility to certain pests (positive) and losses in yield due to selfing as well as troubles caused by cross pollination in addition to contradiction noticed in results on the same problem of various investigators could be (or some of them) limiting factors in any similar programme.

REFERENCES

Abdalla, M.M.F. 1964. Variation of some agronomic characters in different collections of Vicia faba L. M.Sc. Thesis, Fac. of Agric. Cairo University.

Abdalla, M.M.F. 1979. A bibliography of field beans (Vicia faba L.) research in Egypt. ICARDA, Aleppo, Syria.

Abdalla, M.M.F. 1981. Mutation breeding in faba beans. Proc. of Int. Conf. on faba beans, 7-11 March, 1981, Cairo, Egypt (in press).

Abdalla, M.M.F., Morad, M.M. and M. Roushdi. 1976. Some quality characteristics of selections of Vicia faba L. and their bearing upon field bean breeding. Z. Pflanzenzuchty, 77, 72-79.

Abo-Hegazi, A.M.T. 1973. Progress in a breeding programme for high protein quality and quantity through the radiation induced mutations in the main pulses grown in Egypt. Proc. of the 7th Sci. Arab Congr., Cairo, Sept. 1973.

Abo-Hegazi, A.M.T. 1975. Utilization of gamma rays for the improvement of field bean (Vicia faba L.). Studies on seed protein content and its relationsh;hip with other plant characteristics. In : Proc. of 1st Scientific Conf. on the Iraqi Atomic Energy Commission held in Baghdad, Iraq, 7-12 April, 1975, pp. 57-61.

Abo-Hegazi, A.M.T.. Ibrahim, A.A. and Nassib, L. 1978. Environmental effects on seed protein in relation to yield components in field bean. Poljoprivredna Znansvena smotra, 46 (56: 95-104).

Abo-Hegazi, A.M.T. and Ahmed, M.Y.Y. 1978. Attempts towards the understanding of the mechanisms of resistance to Callosobruchus maculatus F. in lines and radiation-induced mutations of field bean, Vicia faba L. Proc. of 4th Conf. Pest Control, NRC, Cairo, pp. 803-811.

Abo-Hegazi, A.M.T. 1979. High protein lines in field beans Vicia faba from a breeding programme using gamma rays. 1. Seed yield and heritability of seed protein. In : Seed Protein Improvement in Cereals and Grain Legumes, IAEA, Vienna, Vol. II, 33-36.

Abo-Hegazi, A.M.T., Tolba, A.E.M., Eweida, M.H.T. El-Agamy, A.I. and Haggag, A.E.A. 1981. Effect of Rhizobium inoculation on some characters of a high protein mutant and its parent field bean Vicia faba L. In : Induced Mutations, a tool in Plant Breeding, IAEA, Vienna, pp. 323-326.

Abo-Hegazi, A.M.T. 1982. Prospects of inducing resistance to Callosobruchus chinensis and Bruchidius incarnatus through mutations. Final report on IAEA Contract No. 2672/RB, FAO/AEA, Vienna (unpublished).

Aykroyd, W.R. and Doughty, J. 1964. Legumes in human nutrition. FAO of the UN, Rome, No. 19, 3.

Ali, A.M.A. 1970. Morphological and cytological studies of field bean, Vicia faba L. M.Sc. thesis, Fac. Agric. Alexandria Univ., Egypt.

Anonymous, 1971. In : Recent Research on the Improvement of Protein Properties of Food and Feed Plants. IARI, New Delhi, India, Res. Bull. New Series, No. 6, p.v.

Atia, Z.M.A. 1981. Effect of gamma radiation on growth promotors and inhibitors in relation to shedding and yield in Vicia faba L. M.Sc. thesis, Cairo Univ., Egypt.

Bond, D.A. 1977. Breeding for zero tannin and protein yield in field bean (Vicia faba L.). In : Protein quality from leguminous crops. A seminar held at Dijon, France, Nov.3-5, 1976. EUR 5686, EN 348-360.

El-Betal, M.A.A. 1979. A study in effect of some agricultural treatments on plant characters and yield of field bean. M.Sc. Thesis, Al-Azhar Univ., Cairo, Egypt.

El-Kady, M.A. 1978. Induced variability of yield and yield components in two Egyptian broad bean cultivars by gamma irradiation. Res. Bull. No. 820, Fac. of Agric. Ain Shams Univ., Cairo, Egypt.

El-Sherbeeny, M.H. 1970. Studies on pollination, fertilization and pod-setting in the field bean and their bearing on breeding the crop. M.Sc. thesis, Cairo Univ.

Griffiths, D.W. and Lawes, D.A.1978. Variation in crude protein content of field beans (Vicia faba) in relation to the possible improvements of the protein content of the crop. Euphytica 27, 487-495.

Hassan, H.F. 1977. Mutation studies on Vicia faba. M.Sc. thesis, Al-Azhar Univ., Cairo, Egypt.

Hassan, H.F, 1982. Hybridization and its effect in improving some agronomic and technological characteristics of field bean (Vicia faba L.) Ph.D. thesis, Al-Azhar Univ., Cairo, Egypt.

Hawtin, G.C. 1981. An overview of breeding methods for the genetic improvement of faba beans. Proc. of Int. Conf. on faba beans, 7-11 March, 1981, Cairo, Egypt, (in press).

Hussein, H.A.S. and Abdalla, M.M.F. 1978. Protein and yield traits of field bean mutants induced with gamma rays, EMS and their combination. In : Seed protein improvement by nuclear techniques. IAEA, Vienna, 253-264.

Hussein, H.A.S. and Abdalla, M.M.F. 1979. Gamma-ray and EMS-induced mutations in Vicia faba L., Evaluation of yield and protein traits of mutants in the M₄ and M₅ generations. In : Seed protein improvement in cereals and grain legumes. Vol. II, IAEA, Vienna, 23-31.

Ibrahim, A.A. 1954. A comparative study of six varieties of beans (Vicia faba L.) with respect to their branching, flowering and fruiting characteristics. M.Sc. thesis, Fac. of Agric., Cairo Univ.

Ibrahim, A.A. 1963. A genetic analysis of some Egyptian and imported varieties of beans (Vicia faba L.). Ph.D. thesis, Cairo, Univ.

Ibrahim, A.A. 1981. Overview of Egyptian National Program. Proc. of Int. Conf. on faba beans, 7-11 March, 1981, Cairo, Egypt. (in press).

Kaser, H.R. and Steiner, A.M. 1983. Subspecific classification of Vicia faba L. by protein and isozyme patterns. FABIS, ICARDA, Aleppo, Syria, No. 7, pp.19-20.

Khalil, S.A. 1984. Screening for disease resistance in faba bean (Vicia faba L.). Seminar presented April 5 at ARC, Giza, Egypt.

Metwally, M.A 1973. Study on the effect of irrigation and fertilization on yield and technological properties in field bean Vicia faba L. M.Sc. thesis, Al-Azhar Univ., Cairo, Egypt.

Mitkees, R.A. and Hassan, H.F. 1983. A diallel cross analysis of some chemical constituents of faba bean, FABIS, ICARDA, Aleppo, Syria, No. 7, pp. 21-22.

Omar, A.M., Selim, A.K.A., Hassanein, S.H. and Abdel-Hafiz, S.M. 1970. Mode of inheritance of protein content and seed weight in broad bean Vicia faba, Ain Shams Univ. Press, Cairo, Res. Bull. No. 626.

Picard, J., Berthelem, P., Duc, G. and Le Guen, J. 1981. Male sterility and future prospects for hybrid varieties in Vicia faba. Proc. of Int. Conf. on faba beans, 7-11 March, 1981. Cairo, Egypt. (in press)

Rabeia, B.M.B. 1977. Effect of some growth regulators on yield and quality of field bean Vicia faba L. M.Sc. thesis, Al-Azhar Univ., Cairo, Egypt.

Saber, H.A. 1980. Study on the effect of some cultural treatments on yield and quality in field bean. M.Sc. thesis, Al-Azhar Univ., Cairo, Egypt.

Shalaby, T.A.M. 1965. Studies on genetic advance under selection in Vicia faba L. M.Sc. thesis, Fac. of Agric. Cairo Univ.

Sjodin, J. 1981. Protein quality and quantity in Vicia faba Proc. of Int. Conf. on faba beans, 7-11 March, 1981, Cairo, Egypt. (in press)

Tolba, A.M. 1980. Studies on characters of gamma irradiated field bean. M.Sc. thesis, Al-Azhar Univ., Cairo, Egypt

Udy, D.C 1971. J. Am. Oil Chemists Soc. Jan. 48 : 1.

Vries, A. Ph. de. 1979. In search of characters to be used for indirect selection on grain and protein yield in Vicia faba L. In : Some current research on Vicia faba in Western Europe. A Seminar held at Bari, Italy, 27-29 April, 1978. EURO 6244, EN. 324-341.

8
Conclusions

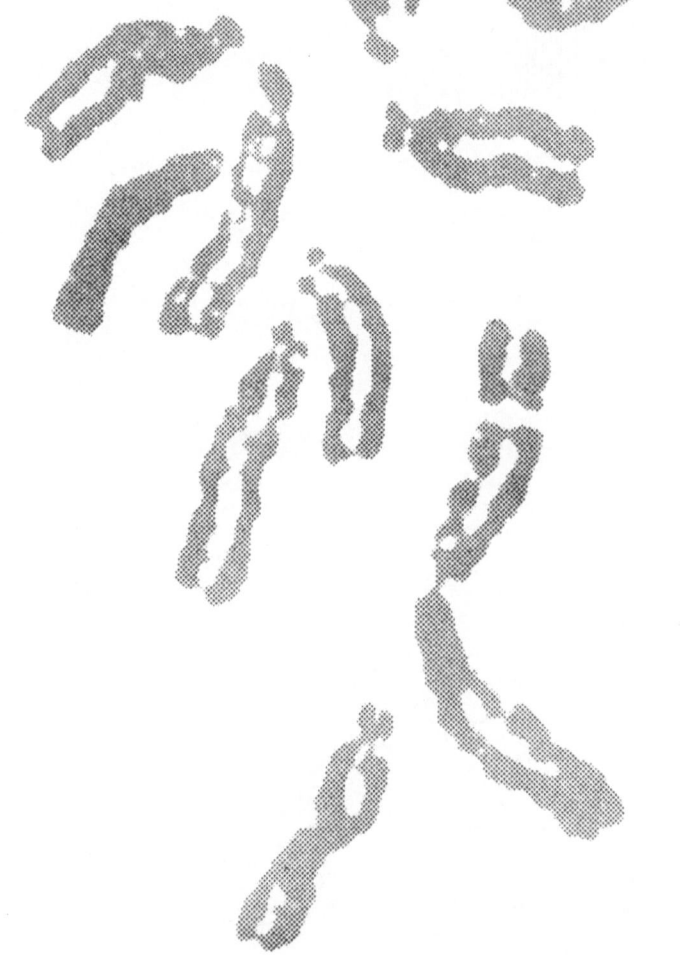

CONCLUSIONS OF THE SECOND INTERNATIONAL VICIA
FABA CYTOGENETICS REVIEW MEETING*

At the concluding session of the seminar, there was a general discussion which led to three resolutions, these being as follows:

1. The recommendations of the Sutton Bonington General Review Meeting (September, 1983) were endorsed and the priority given to cytogenetics noted especially.

2. There should be continuing emphasis in the Vicia faba research programme to establish linkage groups on each chromosome which could then be individually researched.

3. Recognising that tissue culture is of importance to Vicia faba, we wish to support intensive efforts into in vitro techniques in this species.

The value of exchanging workers between laboratories was recognised as were the similar interests of Pisum cytogeneticists who it was felt could usefully be invited to meetings for Vicia faba cytogenetics.

At a future meeting, the emphases were to be for

1) Chromosome research and genetics of Vicia faba
2) Breeding research in Vicia faba
3) Protein metabolism and genetic engineering
4) Germplasm resources of Vicia faba

*The conclusions of the First International Vicia faba Cytogenetics Review Meeting were published in FABIS 6, 19-20. 1983.

9
Appendix

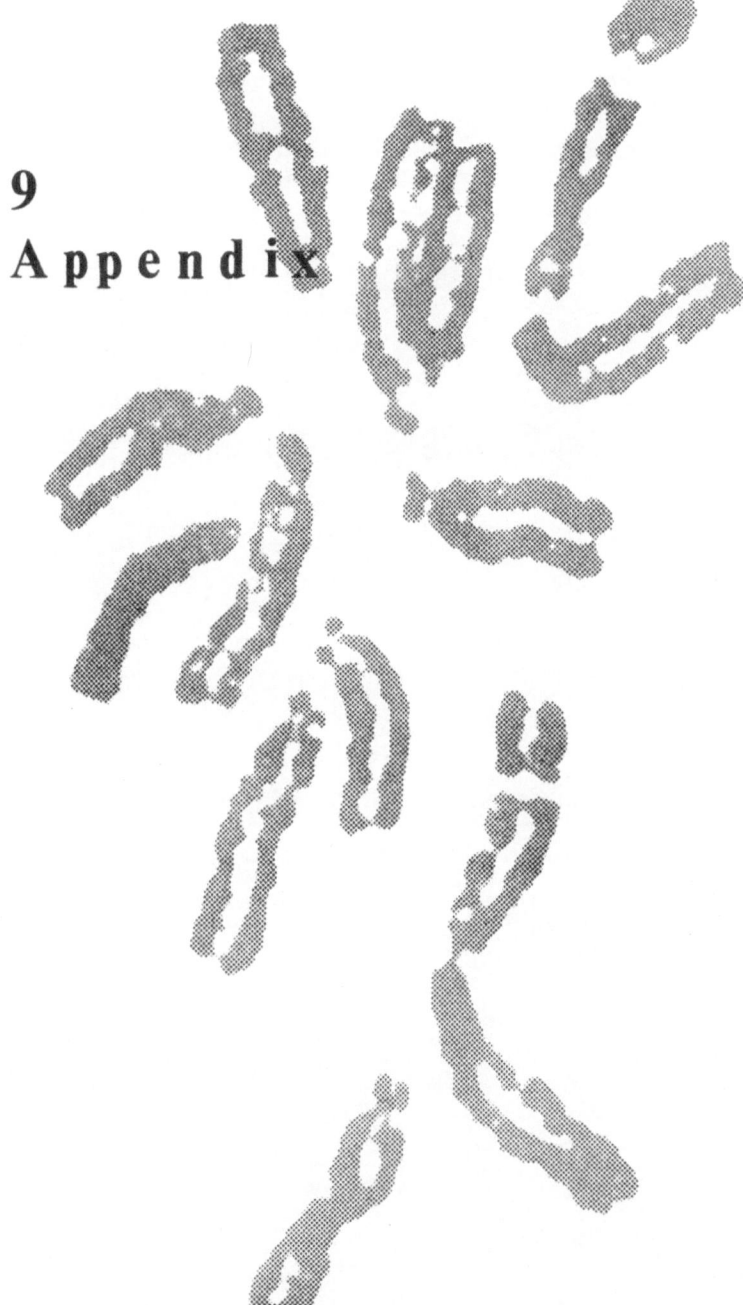

GENETIC VARIATION WITHIN <u>VICIA</u> <u>FABA</u>

(a note on this reprinting)

In 1981, FABIS published as a supplement the following list of genetic variation. By kind permission of ICARDA, the list is reprinted here with some minor amendments. Firstly, the Sjodin (1971) chromosome numbering has been replaced by that due to Michaelis and Rieger (1959, 1968). Secondly, the list of determinate mutants has been augmented. Thirdly, minor typographical errors have been corrected. A more substantial revision is in preparation.

Since the Introduction to the list was written, there has been progress in several directions and some shifts of emphasis and it is appropriate here to make the following points.

1. In several laboratories the range of chromosome aberrations has been systematically extended and their behaviour at meiosis in relation to their use in linkage studies is increasingly receiving interest.

2. The relative ease with which polyteny can be induced in cotyledon nuclei especially in the related genus Pisum suggests that a complementary approach to linkage mapping in <u>Vicia</u> <u>faba</u> could and should be developed.

3. The search for disease resistances especially to <u>Botrytis</u> <u>fabae</u> has made progress and for this reason faba bean populations in the New World have proved surprisingly interesting. Presumably these populations could yield other useful variants beside those for disease resistance and a systematic search there would seem timely.

4. If, as seems likely, some breeders will concentrate their efforts on incorporating closed flower mutants so as to move toward cereal-type breeding programmes, the need for sustainable autofertility becomes more pressing and considerable emphasis should be given to this character.

5. Incorporation into a plant phenotype of either closed flower or determinate habit imposes a yield penalty. Since the theoretical advantages of these alternatives either singly or even in combination are fairly obvious, there is a good case for seeking less deleterious alternatives. Several exist for example for determinate growth but only one appears to have been rigorously assessed by breeders.

Beyond these five points, there seems no reason to modify other aspects of the original document.

REFERENCES

Chapman, G.P. 1981. Genetic variation within Vicia faba. FABIS, No. 3, supplement, pp. 12.

Michaelis, A. and Rieger, R. 1959. Structurheterozygotie bei Vicia faba Zuchter 29, 354-361.

Michaelos, A. and Rieger, R. 1968. On the distribution between chromosomes of chemically induced chromatid aberrations: studies with a new karyotype of V. faba. Mutation Res. 6, 81-92.

The most comprehensive assessment of genetic variation in Vicia faba undertaken hitherto was that of Sirks (1931). Since all the material was lost during the Second World War the most realistic approach for breeders is to accept the value of Sirks' insights, but to recognise that the variants with which they work cannot be unequivocally equated with the similar ones to which Sirks refers. Only presently available variation has practical value to the breeder and where a choice regarding symbols was required preference has been given in this list to current usage.

In assembling the present list the following considerations require mention:

1. **Availability.** This is understood to mean 'known to exist' rather than 'available for distribution'.

2. **Citation.** Journal references were not included since it would have added unustifiably to the length of the list. Where possible I have named individuals that might be contacted via the FABIS mailing list. Alternatively, scientific establishments are indicated.

3. **Authenticity.** Every effort has been made to ensure the accuracy of information included here, but this is both the first attempt to catalogue existing variation and in many cases the list deals with information in need of re-examination. On balance there seems to be a convincing case for each worker to have a list, so that problems can be identified and attacked. In this connection my approach here has been that of editor rather than arbitrator.

4. **Genetic status of variation.** Not all variation can be resolved into single gene units nor should it be assumed that what is most frequently seen is necessarily dominant. The commonplace situation is underlined. Where known or suspected, pleiotrophy is indicated, and that associated with white flowers merits re-examination as to

its extent.

5. **Chromosomes.** Several systems of nomenclature have been proposed and the one adopted here is that due to Sjodin (1971). This author tabulates the corresponding chromosome designations from the various systems.

6. **Other Vicia species.** No reference to other Vicia species has been included since V. faba appears to be genetically isolated with little prospect of this being modified in the near future.

Future priorities.

In the course of editing the following table of variation several priorities emerged of which the following are perhaps the most important. They are offered as recommendations.

1. **Diverse chromosome stocks.** The number available is possibly quite large and includes a wide range of useful aberrations. It is suggested that some attempt is made to make generally available a tester translocation set and to encourage the development of a complete set of stable trisomics.

   By these means a co-ordinated study of linkage could proceed, revealing in the process some information about this species' genetic system.

2. **New variation.** It could be questioned whether searches for yet more ovules per pod or more extreme forms of dwarfism were needed at present. What seems much more important is information about the genetic bases of variation in the pests and diseases of the crop, and the crop's range of response to them. As will be apparent, the list of faba bean maladies is not short, but available genetic knowledge on these pests and diseases is meagre.

3. **Centralisation.** Maintenance of a wide range of genetic stocks is time-consuming and a worker wishing to move on to other subjects should surely have the option of being freed from such a task. A central system that included training of suitable staff to undertake such work might be advantageous. The crop out-pollinates, and assumptions made on the basis of cereal line propagation are inappropriate. This does not seem always to be appreciated.

4. **Publication of new information.** While information will be published in a range of journals regarding new mutants or the revised status of existing ones it would simplify the task of updating the present list

if the essentials were published at the same time in FABIS. This could include a brief description and an entry for each of the columns in the present table, except the last. Ascribing a gene to a chromosome is a longer term process and could be added subsequently.

It is increasingly realised that two or three generations per year of _Vicia faba_ (under properly controlled conditions and using appropriate phenotypes) is a feasible rate of genetic turnover. Added to this, the diversity available and the manipulability of the chromosomes give the possibility of accelerated improvement for this species and an increased understanding of the genetic system.

In addition to the benefits breeders might expect from this catalogue of available variations, it is likely that agronomists, physiologists and processors will be encouraged to explore the implications of having alternatives to the familiar tall leafy indeterminate phenotype.

GENETIC VARIATION WITHIN VICIA FABA

N.B. This list is based on information currently available and is subject to periodic revision.

feature	variation (the commonplace situation is underlined)	locus and dominance relations	origin	chromosome ascription
LEAF	bi- to mulitifoliate during life cycle			
	unifoliate (obligate)	un-a^1	induced, X-irradiation, 8000r. 1958 Sjodin	
	(obligate)	un-a^2	spontaneous, obtained from Gottschalk 1961	
	(obligate)	un-a^5	spontaneous, isolated in breeding material, 1966	
	(obligate)	un-a^6	spontaneous, isolated in breeding material, 1966	
	(obligate)	un-a^7	induced, MMS. 0.05% 1965	
	(obligate)	un-a^8	induced, neutrons, 140 rad. 1966	
	(transnormal)	un-bc^1	spontaneous, found in Bohuslan, Sweden, 1961, Sjodin	
	grey-green colour			
	bluish variant of above			
LEAFLET	larger about 10 cm x 5 cm			
	smaller about 6 cm x 3 cm			
TENDRIL	not more than 2 cm, not subdivided			
	longer than 2 cm (subdivision?)			
STIPULE	small			
	large			
	serrate			
	spotted	sp-a	see floral mutants	
	unspotted	sp-b	(pleiotropy)	

STEM

Character	Symbol	Origin
indeterminate with axillary inflorescences	Ti	
determinate with terminal inflorescence	ti-1	neutrons 35 rd. Sjodin
	ti-2	MMS 0.015% Sjodin
	ti-3	mutagen treatment Nagl
	ti-4	mutagen treatment Steuckardt
semideterminate with terminal inflorescence	ti-5	spontaneous mutant Cubero
	Ti-g	spontaneous mutant Frauen (dominant) long arm c'some V Sjodin
		(A new mutant provisionally here termed ti-6 has been reported by Filipetti.)
tall		
compact (short internodes giving dwarf appearance)	dw-1	from Bond 1964 from variety Compacta
	dw-2	spontaneous Svalof 1970
main stem with one to three side branches		
main stem with many (up to 15) side branches. Associated with terminal flowering.		
anthocyanins present	Rs	
anthocyanins absent giving green stem	rs	Plant Breeding Institute Cambridge. Line 349. Bond (rs condition occurs in Triple white-pleiotropy)
erect		
decumbent		
±prostrate		Dijon collection, France, Picard
basal branching		
high incidence of branches arising from higher nodes		

feature	variation (the commonplace situation is underlined)	locus and dominance relations	origin	chromosome ascription
INFLORESCENCE	axillary	(see ti above)		
	terminal			
	multiflowered (up to 10 or 12)		as in subsp. paucijuga	
	one or two flowered			
	pedicels arising close to base of peduncle			
	pedicels arising more than 2cm from base			
	asynchronous flowering (within inflorescence)			
	synchronous flowering	reported by Gates 1980 (Durham)		
FLOWER	ground colour off-white			
	violet	$dp\text{-}a^1$	EI 0.0031% 1958	
	violet	$dp\text{-}a^2$	X-irradiation, 1956	
	dark brown	$dp\text{-}a^3$	X-irradiation, 1956	
	dark brown	$dp\text{-}a^4$	X-irradiation, 1956	
	dark brown	$dp\text{-}a^5$	X-irradiation, 1956	
	dark brown	$dp\text{-}a^6$	X-irradiation, 1956	
	pink standard, violet wings	$dp\text{-}a^7$	X-irradiation,6000r.1958	
	light pink standard, brown wings	$dp\text{-}a^8$	X-irradiation,8000r.1957	
	pink standard, brown wings	$dp\text{-}a^9$	EI 0.0031% 1959	
	light pink standard, brown wings	$dp\text{-}a^{10}$	EI 0.0062% 1959	
	violet standard, brown wings	$dp\text{-}a^{11}$	EI 0.025% 1959	
	violet	$dp\text{-}a^{12}$	EI 0.025% 1959	
	dark brown	$dp\text{-}a^{13}$	EI 0.025% 1959	
	dark brown	$dp\text{-}a^{14}$	X-irradiation, 7000r. 1959	short arm c'some I
	dark brown	$dp\text{-}a^{15}$	X-irradiation, 7000r. 1959	(satellite c'some)
	violet	$dp\text{-}a^{16}$	X-irradiation, 7000r. 1959	
	dark brown	$dp\text{-}a^{17}$	spontaneous, from Bohuslan, 1961	(Sjodin)
	dark brown	$dp\text{-}a^{18}$	spontaneous, from Bohuslan, 1961	
	violet			

dark brown   dp-a19   neutrons, 132 rad. 1961

pink   dp-a20   spontaneous, obtained from Bond (9311/1/1/1), 1963

dark brown   dp-a21   spontaneous, obtained from Bond (9311/1/2/3), 1963

violet   dp-a22   spontaneous, obtained from Bond (51/3 x 9311), 1963

scarlet   dp-a23   spontaneous, obtained from Bond (C5/2/12/1) 1963

dark brown   dp-a24   spontaneous, obtained from Bond (C8/2/2/1/1/1), 1963

dark brown   dp-a26   spontaneous, obtained from Rowlands (VI 63/8), 1964

dark brown   dp-a27   spontaneous, obtained from Rowlands (CI), 1964

dark brown   dp-a28   spontaneous, obtained from Rowlands (C8), 1964

solid yellow   dp-a29   spontaneous, obtained from Rowlands (IV 63/2), 1964

yellow wing spots   dp-b1   spontaneous, obtained from Rowlands (VI 6309), 1964

greyish brown   dp-a31   gamma rays, 9000 r. 1967

wing petals dark spotted, standard dark striped

unspotted   sp-a1   spontaneous, obtained from Bond, 1963

unspotted   sp-a2   spontaneous, obtained from Rowlands (Ch 170), 1964

unspotted   sp-b1   spontaneous, obtained from Rowlands (Triple White), 1964

unspotted   sp-b2   spontaneous, obtained from Rowlands (VI 6302), 1964

unspotted   sp-b3   spontaneous, obtained from Rowlands (VI 6301), 1964

unspotted   sp-b4   spontaneous, obtained from Picard,1968

flowers 2.5-3.5 cm long
flowers less than 2.5 cm long

feature	variation (the commonplace situation is underlined)	locus and dominance relations	origin	chromosome ascription
	leguminous keel present separated or diverging keel petals exposing the stigma, (associated with 'unifoliate', sometimes)			
	open flower closed flower partially closed flower	cf	spontaneous? Poulsen, 1977	
POLLEN	prolate round triangular round-triangular	Po po-1* po-2 po-3	X-ray,4000r. *see later note on tetraploidy neutrons 140 rad. MMS. 0.01%	
	fertile nuclear genetic male sterility	ms	PBI (An additional nuclear male sterility reported by Picard, Dijon)	
	cytoplasmic male sterility	447 350	PBI INRA	
	restorer	Rf1 Rf2	8.45 PBI LCF PBI	
CARPEL (at flowering)	partial self compatibility self compatible auto fertile			
	four ovuled one or two ovules eight or more ovules		some Ethiopian material some V. faba major selections	

CARPEL (at maturity)

erect (V. faba minor)
pendent (V. faba major)
horizontal

straight
curved

with felty indumentum
without felty indumentum

dehiscent
indehiscent

SEED

about 1 cm long, bolster shaped
down to 0.5 cm long, bolster shaped
up to 2.5cm long, flattened

favism** positive
favism negative or low incidence?

'black' seeded
(sometimes regarded as very
dark brown or dark violet)

some V. faba major cultivars

Triple White (Threefold White) Revoira G.1979

Sc-1	isolated in X-irradiated material,1956
Sc-2	isolated in X-irradiated material,1956
Sc-3	isolated in EI-treated material, 0.125% 1958
Sc-4–Sc-7	isolated in X-irradiated material,1956
Sc-6	Long arm c'some II (Sjodin)
Sc-8	isolated in X-irradiated material,4000r. 1956
Sc-9–Sc-13	isolated in X-irradiated material, 1956
Sc-14	isolated in X-irradiated material,8000r. 1956
Sc-15	isolated in X-irradiated material,4000r. 1956

** among genetically susceptible humans

feature	variation (the commonplace situation is underlined)	locus and dominance relations	origin	chromosome ascription
		Sc-16	isolated in EI-treated material,0.05% 1960	
		Sc-17	isolated in X-irradiated material,1957	
		Sc-18	wild form from Italy, obtained from Gatersleben, 1961	
		Sc-19-Sc-28	wild forms from Balkans, obtained from Gatersleben, 1961	
		Sc-29	Kiemes Schwartze Pferdebohne, obtained from Gatersleben, 1961	
		Sc-30	"Fioletowy", obtained from Warsaw,1961	
		Sc-31	"Feverole de Florence", obtained from Rabat, 1962	
		Sc-32	isolated in EI-treated material,0.025% 1959	
		Sc-33	isolated in EI-treated material,0.05% 1959	
		Sc-34	isolated in EI-treated material,0.1% 1959	
		Sc-35	isolated in EI-treated material,0.2% 1959	
		Sc-36	isolated in X-irradiated material,6000r 1959	
		Sc-37	isolated in EI-treated material,0.0015% 1959	
		Sc-38	isolated in X-irradiated material,8000r 1959	
		Sc-39	isolated in EI-treated material 0.025% 1959	
		Sc-40	obtained from Bot.Garden,Moscow,1959	

Sc-41 isolated in X-irradiated material,6000r. 1967

Sc-42 isolated in MMS-treated material, 0.02% 1967

Sc-43 isolated in MMS-treated material, 0.01% 1967

violet seeded

V-1 isolated in Primus, 1959
V-2 isolated in EI-treated material, 0.05% 1960

V-3 isolated in Primus, 1959
V-4 obtained from Bot.Garden,Moscow,1962
V-5 obtained from Bot.Garden,Moscow,1962
V-6 obtained from Rowlands (AD99),1964
V-7 obtained from Bryssine,Rabat,1964 Polish cultivars

buff seeded
yellow seeded
green seeded

y-1 see Dijon below wild form from China,obtained from Gatersleben, 1961

y-2 otained from Rowlands(Ch193),1964 long arm c'some IV Sjodin

y-3 from Japan through FAO,1965
y-4 obtained from Bond,Line 349,1968
y-5 obtained from Bond,"Staygreen",1968
red seeded
r-1 obtained from Rowlands(AD96),1964
r-2 obtained from Picard(D1434),1967 red spotted - see Dijon below

unspeckled
speckled

unstriped
striped South American collections at Dijon, France. Picard

feature	variation (the commonplace situation is underlined)	locus and dominance relations	origin	chromosome ascription
	unhooked (i.e. without cotyledon bulges) hooked	Ho ho	EI 0.015%, Sjodin	
	tough testa papery testa semi-naked	sn	spontaneous, Poulsen	
	long-narrow hilum 1mm × 4-5mm small hilum about 0.5mm × 3mm	Hi-1 Hi-2 Hi-3 Hi-4		
	round hilum long hilum 1mm × 6mm			
	black hilum colourless hilum	n	Fyfe 1951	
	medium protein(25-30%) low protein to 15% high protein to 45%			
	low oil content high oil content?			
	low sulphur amino acid high sulphur amino acid?			
	high percentage of hard seeds low percentage of hard seeds		cvs Sudanese Triple White, Kambal and Salih	
	short cooking time long cooking time			

ROOT	nodulated non-nodulated?		
	normal rooting depth deep rooting		evidence that certain Mediterranean lines can extract soil moisture from a greater depth than normal
WHOLE PLANT	photoperiod sensitive photoperiod insensitive		winter beans spring beans
	vernalisation sensitive vernalisation insensitive		
	frost susceptible frost resistant		most lines of V.faba U.K. and French winter varieties and many West Asian lines eg. ILB 1813 and ILB 1814 (Syria, ICARDA) cvs. Hudeiba 72,Sudan;Giza 4,Egypt
	highly frost susceptible		
	diploid(2n equals 12) tetraploid		
	normal six bivalent formation asynaptics,inversions translocation, trisomics	po-1	Poulsen and Martin, 1977 a wide and increasing range of aberrations is being developed.
	drought susceptible drought tolerant		
	sensitive to soil salinity less sensitive to soil salinity?		

feature	variation (the commonplace situation is underlined)	locus and dominance relations	origin	chromosome ascription
ORGANISMS WITH WHICH VICIA FABA INTERACTS				
ANGIOSPERMS	Orobanche crenata		line F402 resistant. A.M. Nassib 1979 cv. Express, L.Kasasian 1973	
BACTERIA	Rhizobium leguminosarum		strains shown to vary in nodule effectiveness with particular V. faba materials. Lawes et al., P.B.S., Wales	
FUNGI	Alternaria tenuis Aphanomyces eutiches Ascochyta fabae Botrytis fabae (and B. cinerea - see note)		partial resistance in some ICARDA lines partial resistance in some PBI and ICARDA lines to B. fabae. Some plant populations showed varied field performance. (Ascochyta and Botrytis symptoms apparently can be confused in the field and the role of B. cinerea requires clarification -G.P.C.)	
	Fusarium avenaceum F. fabae F. oxysporum F. solanum Peronospora viciae Rhizoctonia solani Sclerotinia trifoliorum Stemphylium botryosum			

Uromyces fabae

susc frl three loci identified as **Frl,Fr2,Fr3** (resistant)
susc fr2 with susceptible recessives and extra R loci as
susc fr3 yet unidentified
e.g. Ackerperle **Frl Fr2** fr2 fr3 fr3
Erfordia frl frl **Fr2 Fr2** fr3 fr3 **RR**
Maris Bead-1-2. frl frl **Fr2 Fr2 Fr3 Rr**
Conner and Bernier 1980

INSECTS

Acyrthosyphon pisum

red and green strains with differing fecundity
etc. H.J.B. Lowe et al. Univ. Wales, Cardiff
partial resistance var. Rastatt. Miller,PBI

Aphis fabae
Apion aethiops
Apion vorax
Myzus periscae
Sitona hispidulus
Sitona lineatus

NEMATODES

Ditylenchus dipsaci

no known resistant V. faba. Hooper, Rothamsted
Same author reports several races of this
nematode

Heterodera gottingiana

VIRUS

bean leaf roll

Some resistance in Maris Bead. Maxime and Minor
least susceptible. Herz Freyer and Minden
most susceptible. Cockbain, Rothamsted
several known virus strains. Cockbain

bean yellow mosaic (broad bean yellow mosaic)
broad bean mottle
broad bean stain
broad bean wilt
Echtes Ackerbohnen mosaic (broad bean true mosaic)
pea enation mosaic

Additional information is required: for example, about the various larvae found in the stem lumen.

N.B. Only some of the more important pests and diseases have been listed above. A complete list of all known faba bean pests and diseases is needed.

Please send any additions to:

'Vicia faba heritability list'
FABIS
Training and Communications,
ICARDA, P.O. Box 5466,
Aleppo, Syria.

All additions and amendments to the list will be published in the FABIS Newsletter

INDEX